D0248080

2.50

Chapman and Hall Chemistry Textbook Series

A series of short textbooks on modern topics in chemistry whose prime aim is to provide accounts suited to the needs of undergraduate students in chemistry.

CONSULTING EDITORS

R.P. Bell, M.A., Hon. L.L.D., F.R.S., Professor of Chemistry at the University of Stirling

N.N. Greenwood, Ph.D., Sc.D., Professor of Inorganic and Structural Chemistry at the University of Leeds

R.O.C. Norman, M.A., D.Sc., Professor of Chemistry at the University of York

OTHER TITLES IN THE SERIES

Symmetry in Molecules J.M. Hollas

Introduction to Molecular Photochemistry C.H.J. Wells

N.M.R. and Chemistry

AN INTRODUCTION TO NUCLEAR
MAGNETIC RESONANCE SPECTROSCOPY

J. W. Akitt

Lecturer in the School of Chemistry
University of Leeds

LONDON

CHAPMAN AND HALL

First published 1973
by Chapman and Hall Ltd
11 New Fetter Lane, London EC4P 4EE

© *1973 J.W. Akitt*

Set by E. W. C. Wilkins Ltd and
printed in Great Britain by T. & A. Constable, Ltd, Edinburgh

SBN 412 10250 1

Distributed in the U. S. A.
by Halsted Press, a Division
of John Wiley & Sons, Inc., New York

Library of Congress Catalog Card Number 72-12339

Preface

About 20 years have elapsed since chemists started to take an interest in nuclear magnetic resonance spectroscopy. In the intervening period it has proved to be a very powerful and informative branch of spectroscopy, so much so that today most research groups have access to one or more spectrometers and the practising chemist can expect constantly to encounter references to the technique. There is a considerable number of textbooks available on the subject but these are invariably written primarily either for the specialist or for the graduate student who is starting to use the technique in his research. The author has, however, always felt that a place existed for a non-specialist text written for the undergraduate student giving an introduction to the subject which embraced the whole n.m.r. scene and which would serve as a basis for later specialisation in any of the three main branches of chemistry.

With this in mind the book has been written in two sections. The first covers the theory using a straightforward non-mathematical approach which nevertheless introduces some of the most modern descriptions of the various phenomena. The text is illustrated by specific examples where necessary. The second section is devoted to showing how the technique is used and gives some more complex examples illustrating for instance its use for structure determination and for measurements of reaction rates and mechanisms. A few problems have been included but the main purpose of the book is to demonstrate the many and varied present uses of n.m.r. rather than to teach the student how to analyse a spectrum in detail. This is done best if it is done concurrently with a students own research.

I am indebted to Dr K. Crosbie, Professor N.N. Greenwood, Dr B. Mann, and to Professor D.H. Whiffen who read and criticised the manuscript and to many former colleagues at Newcastle-upon-Tyne for encouragement and for some of the examples used in the text. I also give grateful acknowledgement to Varian Associates Ltd for permission to reproduced the spectra in Figs. 15, 17, 26, 27, 66 and 75 and to Bruker-Spectrospin Ltd for permission to reproduce the spectra in Figs. 49, 50, 63, 64 and 75.

J.W.A.

Leeds,
January, 1972

Contents

CONTENTS

Part I The theory of nuclear magnetic resonance spectroscopy

The theory of nuclear magnetization 1

1.1 The properties of the nucleus of an atom

The chemist normally thinks of the atomic nucleus as possessing only mass and charge and is concerned more with the interactions of the electrons which surround the nucleus, neutralise its charge and give rise to the chemical properties of the atom. Nuclei however possess several other properties which are of importance to chemistry and to understand how we use them it is necessary to know something more about them.

Nuclei of certain isotopes possess intrinsic angular momentum or spin, of total magnitude $\hbar[I(I + 1)]^{1/2}$. The largest measurable component of this angular moment is $I\hbar$, where I is the nuclear spin quantum number and $\hbar$ is the reduced Plancks' constant, $h/2\pi$. I may have integral or half integral values (0, 1/2, 1, 3/2 ....), the actual value depending upon the isotope. If $I = 0$ the nucleus has no angular momentum. Since I is quantised, several discrete values of angular momentum may be observable and their magnitudes are given by $\hbar m$, where the quantum number m can take the values $I, I - 1$, $I - 2 .... - I$. There are thus $2I + 1$ equally spaced spin states of a nucleus with angular momentum quantum number I.

A nucleus with spin also has an associated magnetic moment μ. We can naively consider this as arising from the effect of the spinning nuclear charge which at its periphery forms a current loop. We define the components of μ associated with the different spin states as $m\mu/I$, so that μ also has $2I + 1$ components. In the absence of an external magnetic field the spin states all possess the same potential energy, but take different values if a field is applied. The origin of the n.m.r. technique lies in these energy differences, though we must defer further discussion of this until we have defined some other basic nuclear properties.

The magnetic moment and angular momentum behave as if they were parallel or antiparallel vectors. It is convenient to define a ratio between them which is called the magnetogyric ratio, γ.

$$\gamma = \frac{2\pi}{h} \cdot \frac{\mu}{I} = \frac{\mu}{I\hbar} \tag{1.1}$$

γ has a characteristic value for each magnetically active nucleus and is positive for parallel and negative for antiparallel vectors.

If $I > 1/2$ the nucleus possesses in addition an electric quadrupole moment, Q. This means that the distribution of charge in the nucleus is non-spherical and that it can interact with electric field gradients arising from the electric charge distribution in the molecule. This interaction provides a means by which the nucleus can exchange energy with the molecule in which it is situated and affects certain n.m.r. spectra profoundly.

Some nuclei have $I = 0$. Important examples are the major isotopes ^{12}C and ^{16}O which are both magnetically inactive — a fact which leads to considerable simplification of the spectra of organic molecules. It is instructive to consider the meaning of zero spin. Such nuclei are free to rotate in the classical sense and so form a current loop but have no associated magnetic moment. We must not confuse the idea of quantum mechanical 'spin' with classical rotation however. The nucleons, that is the particles such as neutrons and protons which make up the nucleus, possess intrinsic spin in the same way as do electrons in atoms. Nucleons of opposite spin can pair, just as do electrons, though they can only pair with nucleons of the same kind. Thus in a nucleus with even numbers of both protons and neutrons all the spins are paired and $I = 0$. If there are odd numbers of either or of both, then the spin is non-zero, though its actual value depends upon orbital type internucleon interactions. Thus we build up a picture of the nucleus in which the different resolved angular momenta in a magnetic field imply different nucleon arrangements within the nucleus, the number of spin states depending upon the number of possible arrangements. If we add to this picture the concept that s bonding electrons have finite charge density within the nucleus and become partly nucleon in character, then we can see that these spin states might be perturbed by the hybridisation of the bonding electrons and that information derived from the nuclear states might lead indirectly to information about the electronic system and its chemistry.

The properties of the most important magnetic isotopes of each element are summarised in Fig. 1, which is set out as a periodic table. The importance of elements for n.m.r. work is indicated by heaviness of type, while the few elements which have effectively no magnetically active nuclei are shown in small type. Each panel of the table contains the spectrometer frequency for a 14 kilogauss magnet, the isotope number of the major active isotope, the spin quantum number I, the nuclear quadrupole moment Q where $I > 1/2$, the natural abundance rounded off to two significant figures, and a relative sensitivity figure which is used for comparing signal areas obtainable from equal numbers of nuclei in a given magnetic field with ^1H taken as unit sensitivity. The resonance heights of nuclei giving resonances of the same width will be in the ratio of these sensitivity figures. If more than one isotope of an element has found use in n.m.r. then this is shown by an +, except in the case of deuterium, ^2H, sometimes written, D, which is important enough to warrant a separate panel of its own.

The usefulness of a nucleus depends upon its natural abundance, sensitivity, and of course upon the chemical importance of the atom it characterises. For this reason most effort has been expended on only a few nuclei, of which the proton (^1H) is the most important*, since it is a constituent of the majority of organic and of many inorganic compounds and gives access to the physical study of many systems. A considerable number of results have also been obtained on boron (^{11}B) which have proved of assistance in the study of boron hydrides, on fluorine (^{19}F) in the now vast array of fluoro organics and inorganics and on phosphorus (^{31}P) in its many compounds. Recently, because carbon is an integral part of all organic molecules and despite its low abundance, considerable effort has been put into ^{13}C spectroscopy also.[†] Note in addition however the large number of elements which are potentially accessible to nuclear magnetic resonance study. The group I and group IIIb metals, nitrogen, oxygen, the halides and xenon have all been given considerable attention already and much remains to be discovered both with these and with other nuclei.

---

* The term proton is commonly used by n.m.r. spectroscopists when discussing the nucleus of neutral hydrogen.

† Since this was written the rapid advances made in Fourier Transform pulse spectrometry (see p 117) have brought ^{13}C into prominence and much new information about the motion of organic molecules is being obtained by comparing the results of ^1H and ^{13}C spectroscopy.

	^{7}Li	^{9}Be		^{1}H	^{2}H	^{3}He
Frequency MHz at 14 kG	23·3	8·4		60·0	9·2	45·6
Spin I	$\frac{3}{2}$	$\frac{3}{2}$		$\frac{1}{2}$	1	$\frac{1}{2}$
Q	−0·040	0·029		−	0·0028	−
Abundance %	93 *‡	100		100	0·015	10^{-7}
Rel. sensitivity	0·29	0·014		1	0·0097	0·44

	^{23}Na	^{25}Mg
Frequency MHz at 14 kG	15·9	3·7
Spin I	$\frac{3}{2}$	$\frac{5}{2}$
Q	0·11	0·22
Abundance %	100	10
Rel. sensitivity	0·093	0·027

	^{39}K	^{43}Ca	^{45}Sc	^{47}Ti	^{51}V	^{53}Cr	^{55}Mn	^{57}Fe
Frequency MHz at 14 kG	2·8	4·0	14·6	3·4	15·8	3·4	14·8	1·9
Spin I	$\frac{3}{2}$	$\frac{7}{2}$	$\frac{7}{2}$	$\frac{5}{2}$	$\frac{7}{2}$	$\frac{3}{2}$	$\frac{5}{2}$	$\frac{1}{2}$
Q	0·09	?	−0·22	0·29	0·2	−0·03	0·35	−
Abundance %	93	0·13	100	7·8	100	9·54	100	2·2
Rel. sensitivity	0·0005	0·064	0·30	0·002	0·38	0·0001	0·18	0·00003

	^{87}Rb	^{87}Sr	^{89}Y	^{91}Zr	^{93}Nb	^{95}Mo	Tc	^{101}Ru
Frequency MHz at 14 kG	19·6	2·6	2·9	5·6	14·7	3·8		3·0
Spin I	$\frac{3}{2}$	$\frac{9}{2}$	$\frac{1}{2}$	$\frac{5}{2}$	$\frac{9}{2}$	$\frac{5}{2}$		$\frac{5}{2}$
Q	0·14	0·36	−	?	−0·2	?		?
Abundance %	27	7	100	11	100	16		17
Rel. sensitivity	0·18	0·0027	0·00012	0·009	0·48	0·0032		0·001

	^{133}Cs	^{137}Ba	^{139}La	^{177}Hf	^{181}Ta	^{183}W	^{187}Re	^{189}Os
Frequency MHz at 14 kG	7·9	6·7	8·5	1·8	7·2	2·5	13·6	4·6
Spin I	$\frac{7}{2}$	$\frac{3}{2}$	$\frac{7}{2}$	$\frac{7}{2}$	$\frac{7}{2}$	$\frac{1}{2}$	$\frac{5}{2}$	$\frac{3}{2}$
Q	−0·003	0·28	0·23	3	3·9	−	2·6	0·8
Abundance %	100	11	100	18	100	14	63	16
Rel. sensitivity	0·047	0·007	0·059	0·00064	0·036	0·00007	0·14	0·0022

	Ce	^{141}Pr	^{143}Nd	Pm	^{147}Sm
Frequency MHz at 14 kG		16·8	3·8		2·1
Spin I		$\frac{5}{2}$	$\frac{7}{2}$		$\frac{7}{2}$
Q		−0·059	−0·482		−0·208
Abundance %		100	12		15
Rel. sensitivity		0·26	0·0028		0·00088

* Of variable abundance either naturally or due to commercial separation

FIGURE 1

Table of main naturally occurring magnetically active isotopes. The table shows only one nucleus for each element, but it should be borne in mind that many elements have several magnetically active isotopes. Some of importance are ^{6}Li, ^{10}B, ^{15}N, ^{37}Cl, ^{79}Br, ^{115}Sn, ^{131}Xe, ^{201}Hg and ^{204}Tl. The n.m.r. frequency in a magnetic field of 14 kG (14 T) is given to the nearest 0·1 MHz. This figure is proportional

Isotope	Frequency	Spin I	Q	Abundance	Sensitivity
^{11}B	19·25	$\frac{3}{2}$	0·036	80*‡	0·17
^{13}C	15·1	$\frac{1}{2}$	–	1·1*	0·016
^{14}N	4·3	1	0·011	99·6‡	0·001
^{17}O	8·1	$\frac{5}{2}$	-0·030	0·037*	0·029
^{19}F	56·4	$\frac{1}{2}$	–	100	0·83
^{21}Ne	4·7	$\frac{3}{2}$	?	0·26	0·0025
^{27}Al	15·6	$\frac{5}{2}$	0·149	100	0·206
^{29}Si	11·9	$\frac{1}{2}$	–	4·7*	0·0078
^{31}P	24·3	$\frac{1}{2}$	–	100	0·066
^{33}S	4·6	$\frac{3}{2}$	-0·10	0·76*	0·0023
^{35}Cl	5·9	$\frac{3}{2}$	-0·079	75·5‡	0·0047
Ar					
^{59}Co	14·2	$\frac{7}{2}$	0·40	100	0·28
^{61}Ni	5·3	$\frac{3}{2}$	?	1?	0·0035
^{63}Cu	15·9	$\frac{3}{2}$	-0·16	69*†	0·093
^{67}Zn	3·8	$\frac{5}{2}$	0·16	4·1	0·0029
^{71}Ga	18·3	$\frac{3}{2}$	0·12	40‡	0·11
^{73}Ge	2·1	$\frac{9}{2}$	0·2	7·6	0·0014
^{75}As	10·3	$\frac{3}{2}$	0·3	100	0·023
^{77}Se	11·4	$\frac{1}{2}$	–	7·6	0·0069
^{81}Br	16·2	$\frac{3}{2}$	0·28	49‡	0·099
^{83}Kr	2·3	$\frac{9}{2}$	0·25	11·5	0·0019
^{103}Rh	18·9	$\frac{1}{2}$	–	100	0·000031
^{105}Pd	2·5	$\frac{5}{2}$	?	22	0·00028
^{109}Ag	2·8	$\frac{1}{2}$	–	49	0·0001
^{111}Cd	12·7	$\frac{1}{2}$	..	13	0·0095
^{115}In	13·1	$\frac{9}{2}$	1·161	96	0·35
^{119}Sn	22·4	$\frac{1}{2}$	–	8·6†	0·052
^{121}Sb	14·4	$\frac{5}{2}$	-0·5	57	0·16
^{125}Te	18·9	$\frac{1}{2}$	–	7	0·032
^{127}I	13·0	$\frac{5}{2}$	-0·70	100	0·094
^{129}Xe	16·6	$\frac{1}{2}$	–	26‡	0·021
^{193}Ir	1·2	$\frac{3}{2}$	1·5	62	0·000042
^{195}Pt	12·9	$\frac{1}{2}$	–	34	0·0099
^{197}Au	1·0	$\frac{3}{2}$	0·60	100	0·000025
^{199}Hg	10·7	$\frac{1}{2}$	–	17‡	0·0057
^{205}Tl	34·6	$\frac{1}{2}$	–	70‡	0·19
^{207}Pb	12·6	$\frac{1}{2}$	–	21*	0·0091
^{209}Bi	9·6	$\frac{9}{2}$	-0·34	100	0·14
Po					
At					
Rn					
^{151}Eu	14·8	$\frac{5}{2}$	0·9	48	0·18
^{157}Gd	2·4	$\frac{3}{2}$	1·0	15	0·00033
^{159}Tb	10·9	$\frac{3}{2}$	1·3	100	0·03
^{163}Dy	2·3	$\frac{5}{2}$	1·6	25	0·00064
^{165}Ho	10·2	$\frac{7}{2}$	2·82	100	0·10
^{167}Er	1·5	$\frac{7}{2}$	2·83	23	0·00031
^{169}Tm	4·9	$\frac{1}{2}$	–	100	0·00055
^{171}Yb	10·6	$\frac{1}{2}$	–	14	0·0042
^{175}Lu	6·8	$\frac{7}{2}$	5·9	97	0·049

to the magnetogyric ratio γ. Spin quantum number I, quadrupole moment Q in units of $e \times 10^{-24}$ cm^2, approximate abundance and sensitivity relative to the proton are also given. For comparison the free electron has spin $\frac{1}{2}$ and a relative sensitivity of $2 \cdot 8 \times 10^8$ since its magnetic moment is 660 times larger than that of the proton. Its resonant frequency would be $\sim 39\ 200$ MHz in the same field, though in practice lower frequencies are used

1.2   The nucleus in a magnetic field

If we place a nucleus in a magnetic field $\mathbf{B}_0$ it can take up $2I + 1$ orientations in the field, each one at a particular angle θ to the field direction and associated with a different potential energy. The energy of a nucleus of magnetic moment μ in field $\mathbf{B}_0$ is $-\mu_z \mathbf{B}_0$, where μ_z is the component of μ in the field direction. The energy of the various spin states is then:

$$-\frac{m\mu}{I}\mathbf{B}_0 \quad \text{or individually} -\mu\mathbf{B}_0, -\frac{I-1}{I}\mu\mathbf{B}_0, -\frac{I-2}{I}\mu\mathbf{B}_0, \text{ etc.}$$

The energy separation between the levels is constant and equals $\mu\mathbf{B}_0/I$. This is shown diagrammatically in Fig. 2 for a nucleus with $I = 1$, and positive magnetogyric ratio. The value of m changes sign as it is altered from I to $-I$ and accordingly the contribution of the magnetic moment to total nuclear energy can be either positive or negative, the energy being increased when m is positive. The energy is decreased if the nuclear magnetic vectors have a component aligned with the applied field in the classical sense. An increase in energy corresponds to aligning the vectors in opposition to the field. Quantum mechanics thus predicts a non-classical situation which can only arise because of the existence of discrete energy states with the high energy states indefinitely stable.

In common with other spectral phenomena, the presence of a series of states of differing energy in an atomic system provides a situation where interaction can take place with electromagnetic radiation of the correct frequency and cause transitions between the energy states. The frequency is obtained from the Bohr relation, namely:

$$h\nu = \Delta E, \text{ the energy separation}$$
$$\text{For n.m.r.} \quad h\nu = \mu\mathbf{B}_0/I$$

In this case the transition for any nuclear isotope occurs at a single frequency since all the energy separations are equal and transitions are only allowed between adjacent levels (i.e. the selection rule $\Delta m = \pm 1$ operates). The frequency relation is normally written in terms of the magnetogyric ratio (1.1) giving:

$$\nu = \gamma\mathbf{B}_0/2\pi \tag{1.2}$$

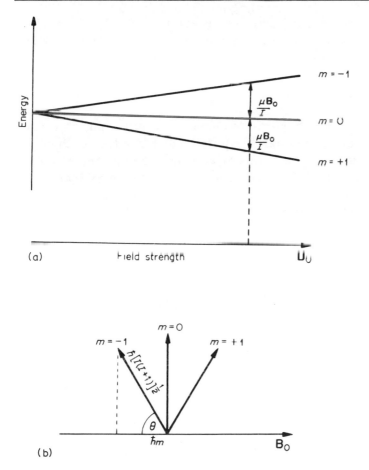

FIGURE 2

(a) The nuclear spin energy for a single nucleus with $I = 1$ (e.g. ^{14}N) plotted as a function of magnetic field B_O. The two degenerate transitions are shown for a particular value of B_O. (b) The alignment of the nuclear vectors relative to B_O which correspond to each value of m. The vector length is $\hbar [I(I + 1)]^{1/2}$ and its z component is $\hbar m$ whence $\cos\theta = m/[I(I + 1)]^{1/2}$

Thus the nucleus can interact with radiation whose frequency depends only upon the applied magnetic field and the nature of the nucleus. Magnetic resonance spectroscopy is unique in that we can choose our spectrometer frequency at will, though within the limitation of available magnetic fields.

9

The values of γ are such that for practical magnets the frequency for nuclei lies in the radio range between a present maximum of 300 megahertz (MHz) and a minimum of a few kilohertz (kHz).

1.3   The source of the n.m.r. signal

The low frequency of nuclear magnetic resonance absorption indicates that the energy separation of the spin states is quite small. Since the nuclei in each of the states are in equilibrium, this suggests that the numbers in the different spin states will not differ greatly. If there is a Boltzman distribution among the spin states then we can expect more nuclei to reside in the lowest energy states. For a system of spin 1/2 nuclei a Boltzman distribution would give:

$$N_h/N_l = \exp(-\Delta E/kT)$$

where N_h and N_l are the numbers of nuclei in the high and low energy states respectively, ΔE is the energy separation, k is the Boltzman constant, and T is the absolute temperature. ΔE is given above as $\mu \mathbf{B}_0/I$ which for $I = 1/2$, equals $2\mu \mathbf{B}_0$. Thus:

$$N_h/N_l = \exp(-2\mu \mathbf{B}_0/kT)$$

which since $N_h \approx N_l$ can be simplified to:

$$N_h/N_l = 1 - 2\mu \mathbf{B}_0/kT$$

For hydrogen nuclei in a field of 14 000 gauss (14 kG)

$$2\mu \mathbf{B}_0/kT \approx 10^{-5}$$

The excess population in the low energy state is thus extremely small. As far as the overall nuclear magnetism of the sample is concerned the effect of all the nuclei in the high energy state will be cancelled by opposing nuclei in the low energy state, only the small excess number with low energy being able to give rise to an observable external magnetic effect, or apparently to absorb radiation.

If $I > 1/2$ one obtains a similar though more complex picture since the excess low energy nuclei do not all have the same value of $m\mu/I$.

It will be noticed that the size of the excess low energy population is proportional to $\mathbf{B}_0$. For this reason the magnetic effect of the nuclei and

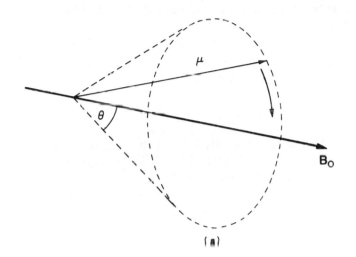

(a)

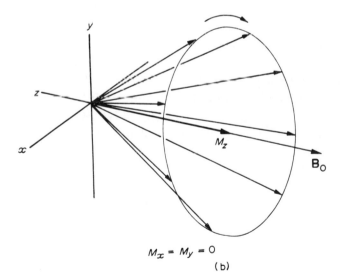

$M_x = M_y = 0$

(b)

FIGURE 3

Freely precessing nuclei in a magnetic field B_o. Larmor precession of (a) a single nucleus (b) The excess low energy nuclei in a sample. The nuclear vectors can be regarded as being spread evenly over a conical surface. They arise from different atoms but are drawn with the same origin.

11

therefore the intensity of their n.m.r. signal increases as the strength of the magnetic field is increased.

We have so far built up a picture of the nuclei in a sample polarised with or against the magnetic field and lying at an angle θ to it. The total angular momentum (i.e. the length of the vectors of Fig. 2) is $[I(I + 1)]^{1/2}$ and the angle θ is then given by:

$$\cos \theta = m/[I(I + 1)]^{1/2}$$

This angle can also be calculated classically by considering the motion of a magnet of moment μ in an applied magnetic field. It is found that the magnet axis becomes inclined to the field axis and wobbles or precesses around it. The magnet thus describes a conical surface around the field axis. The half apex angle of the cone is equal to θ and the angular velocity around the cone is $\gamma \mathbf{B_0}$ so that the frequency of complete rotations is $\gamma \mathbf{B_0}/2\pi$, the nuclear resonant frequency (Fig. 3a). This precession is known as the Larmor precession.

For an assembly of nuclei with $I = 1/2$ there are two such precession cones, one for nuclei with $m = + 1/2$ and one for $m = - 1/2$ and pointing in opposite directions. It is usual however to consider only the precession cone of the excess low energy nuclei, and this is shown in Fig. 3b, which represents them as spread evenly over a conical surface and all rotating with the same angular velocity around the magnetic field axis which is made the z axis. Since the excess low energy nuclear spins all have components along the z axis pointing in the same direction, they add to give net magnetization M_z along the z axis. Individual nuclei also have a component μ_{xy} transverse to the field axis in the xy plane. However, because they are arranged evenly around the z axis, these components all average to zero, i.e. $M_x = M_y = 0$. The magnetism of the system is static and gives rise to no external effects other than a very small, usually undetectable, nuclear paramagnetism due to M_z.

In order to detect a nuclear resonance we have to perturb the system. This is done by applying a sinusoidally oscillating magnetic field along the x axis with frequency $\gamma \mathbf{B_0}/2\pi$ (Fig. 4). This can be thought of as stimulating both absorption and emission of energy by the spin system (i.e. as stimulating upward and downward spin transitions), but resulting in net absorption of energy, since more spins are in the low energy state and are available to be promoted to the high energy state.

Classically we can analyse the oscillating magnetic field into a super-position of two magnetic vectors rotating in opposite directions. These add at different instants of time to give a zero, positive or negative resultant (Fig. 4). The vector $\mathbf{B}_1$ which is rotating in the same sense as the nuclei is

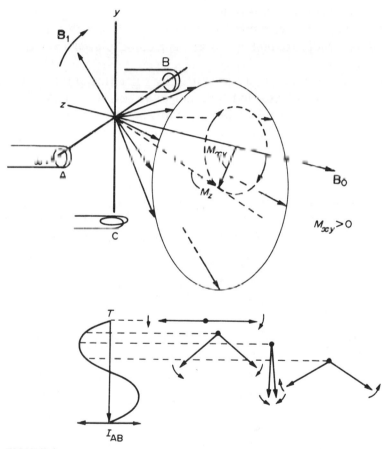

FIGURE 4

If a rotating magnetic vector B_1 with the same angular velocity as the nuclei is now added to the system, the nuclei tend to precess around B_1 and this causes the cone of vectors to tip and wobble at the nuclear precession frequency. The resulting rotating vector M_{xy} in the xy plane can induce a current in coil C (which is normally placed in the xz plane but is shown below this in the figure for clarity). The lower diagram shows how the vector B_1 can be generated by rf currents I flowing in coils AB. The resulting oscillating field can be resolved into two equal vectors rotating in opposite senses.

13

stationary relative to them, since we have arranged that it should have the same angular velocity. The nuclei thus tend to precess around B_1 also and the precession cone axis is displaced from the main field axis. Since the nuclear moments do not now all have the same component in the xy plane they no longer cancel in the xy direction and there is resultant magnetization M_{xy} transverse to the main field and rotating at the nuclear precession frequency. The magnetism of the system is no longer static and the rotating vector M_{xy} will induce a radio frequency current in a coil C placed around the sample.

The signal is weak and could easily be swamped by currents induced in C by the field due to coils A and B, so that we arrange the axis of coils A and B to be at 90° to that of C where they will not couple any energy into C. The output from C is thus wholly our nuclear resonance signal.

If the frequency of the current applied to coils A and B is not equal to $\gamma B_0/2\pi$ then the rotating B_1 vector is not stationary relative to the nuclei, the precession around B_1 is always changing direction and M_{xy} can never become appreciable. If we sweep the frequency of B_1 through a range which includes the nuclear precession frequency we thus will see a single, sharp signal where the nuclei are brought into resonance by the applied oscillating field.

1.4   A basic n.m.r. spectrometer

We are now in a position to understand the principles underlying the construction of an n.m.r. spectrometer. The object is to measure the position of a nuclear resonance and this is done by observing with either a fixed field B_0 and sweeping the frequency of B_1 through resonance or by using a fixed B_1 frequency and sweeping the field and nuclear frequency through resonance. Both types are encountered in practice. The instrument (Fig. 5) comprises a strong, highly stable magnet in whose gap the sample is placed and surrounded by transmitter (AB) and receiver coils C. Either a permanent or an electromagnet may be used. Magnet stability is ensured in the first case by placing the magnet in an isothermal enclosure whose temperature is controlled so as not to drift more than 10^{-4} degrees per hour, while in the second case the magnet energising coils are fed from a highly stable power supply which

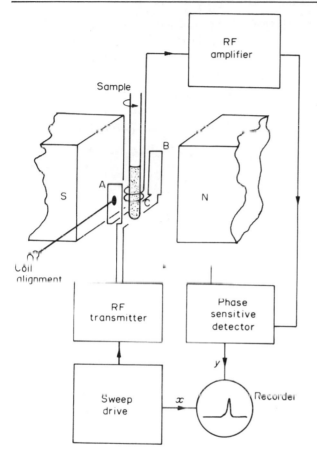

FIGURE 5
A basic n.m.r. spectrometer

detects field variations in an auxiliary coil and makes the required current corrections. The field in the gap inevitably varies throughout the sample volume and the signal is broadened so that a further set of small coils, known as shim coils, is placed around the sample in order to counteract these variations and render the field homogeneous. The shim coils are not shown. Remaining inhomogeneities are minimised by spinning the tube about its long axis so that the sample molecules experience average fields. Very

15

narrow signals and excellent resolution of the resonances is obtained in this way. The B_1 field is produced by a stable, crystal-controlled transmitter and the resonance signal is amplified and then detected in a device which compares it with the input B_1 signal and so produces an output containing both amplitude and phase information. This output is fed to the y axis of a recorder or oscilloscope and the spectrum is developed by sweeping simultaneously the recorder x axis and either field or frequency, as appropriate.

The magnetic field at the nucleus   2

2.1   Effects due to the molecule

So far we have shown that a single isotope gives rise to a single nuclear magnetic resonance in an applied magnetic field. This really would be of little interest to the chemist except for the fact that the magnetic field at the nucleus is never equal to the applied field, but depends in many ways upon the structure of the molecule in which the atom carrying the nucleus resides.

The most obvious source of perturbation of the field is that which occurs directly through space due to nuclear magnets in other atoms in the molecule. In solids this interaction results in considerable broadening of the resonance which obscures much information. In liquids on the other hand, where the molecules are rotating rapidly and randomly, these fields are averaged to zero, and the lines are narrow and show much structure. For this reason chemists are concerned primarily with liquid samples.

Since the magnetic nuclei do not perturb the field at the nucleus we have therefore to consider the effect that the electrons in the molecule may have. We will concern ourselves only with diamagnetic molecules at this stage and will defer till later discussion of paramagnetic molecules possessing an unpaired electron. When an atom or molecule is placed in a magnetic field the field induces motion of the electron cloud such that a secondary magnetic field is set up. We can think of the electrons as forming a current loop as in Fig. 6 centred on a positively charged atomic nucleus. The secondary field produced by this current loop opposes the main field at the nucleus and so reduces the nuclear frequency. The magnitude of the electronic current is proportional to B_0 and we say that the nucleus is screened (or shielded) from the applied field by its electrons. This concept is introduced

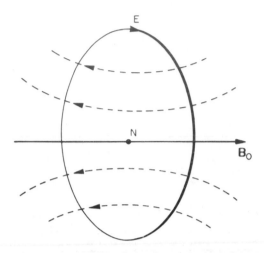

FIGURE 6
The motion of the electronic cloud E around
the nucleus N gives rise to a magnetic field,
shown by dashed lines, which opposes B_0 at
the nucleus.

into equation (1.2) relating field and nuclear frequency by the inclusion of
a screening constant σ.

$$\nu = \frac{\gamma B_0}{2\pi}\,(1 - \sigma) \qquad (2.1)$$

σ is a small dimensionless quantity and is usually recorded in parts per
million, ppm. The screening effect is related to the mechanism which gives
rise to the diamagnetism of materials and is called diamagnetic screening.

The magnitude of the effect also depends upon the density of electrons
in the current loop. This is a maximum for a free atom where the electrons
can circulate freely, but in a molecule the free circulation around an individual nucleus is hindered by the bonding and by the presence of other
positive centres, so that the screening is reduced and the nuclear frequency
increased. Since this mechanism reduces the diamagnetic screening it is
known as a paramagnetic effect. This is unfortunately a misleading term and
it must be emphasised that it does not imply the presence of unparied electrons. As used here it merely indicates that there are two contributions to
σ, the diamagnetic term σ_d and the paramagnetic term σ_p and that these
are opposite in sign. Thus:

$$\sigma = \sigma_d + \sigma_p \qquad (2.2)$$

Since the magnitude of σ_d depends upon the density of circulating electrons, it is common to find in the literature discussion of the effect of inductive electron drifts on the screening of nuclei. The screening of protons in organic molecules for instance depends markedly on the substituents, and good linear correlations have been found between screening constants and substituent electronegativity, thus supporting the presence of an inductive effect. Currently however it is believed that most of these variations originate in the long range effects to be described below and that the contribution of inductive effects is small at least for σ bonded systems.

The magnitude of the paramagnetic contribution σ_p is zero for ions with spherically symmetrical s states but is substantial for atoms involved in chemical bonding. It is determined by several factors. (i) The inverse of the energy separations ΔE between ground and excited electronic states of the molecule. This means that correlations are found between screening constants and the frequency of absorption lines in the visible and ultra violet (ii) The relative electron densities in the various p orbitals involved in bonding, i.e. upon the degree of asymmetry in electron distribution near the nucleus. (iii) The value of $\langle 1/r^3 \rangle$, the average inverse cube distance from the nucleus to the orbitals concerned. In the case of hydrogen, for which there are few electrons to contribute to the screening, and for which ΔE is large, σ_d and σ_p are both small and we observe only a small change in σ among its compounds, most of which fall within a range of 20×10^{-6} or 20 ppm. In the case of elements of higher atomic number, ΔE tends to be smaller and more electrons are present, so that, while both σ_d and σ_p increase, σ_p increases disproportionately and dominates the screening. Thus changes in σ_p probably account for a major part of the screening changes observed for boron in its compounds (a range of 140 ppm), and σ_p almost certainly predominates for fluorine where the range is 1000 ppm, or for thallium where it is 4800 ppm. The changes observed for the proton are thus unusually small.

The observable changes in screening do not increase continuously with atomic number but exhibit a periodicity, being a minimum at the top and a maximum at the bottom of a group of the periodic table then falling again to a new though larger minimum at the top of the next group. This is a consequence of the $\langle 1/r^3 \rangle$ term which exhibits similar periodicity.

The changes observed in screening among the compounds of one element

19

depend primarily upon factors (i) and (ii) above, which means that they are determined by such factors as bond angles and bond order.

The effect of bond angles upon screening is well demonstrated by the phosphorus resonances of some phosphines. ^{31}P screening decreases by 335 ppm between phosphine, PH_3, with bond angles of 93.7°, and trifluoro-phosphine, PF_3, with bond angles of 104°. More strikingly, the screening decreases by 125 ppm between trimethyl phosphine, PMe_3, and tri-t-butyl phosphine, PBu_3^t, due to the increase in bond angle caused by steric crowding in the t-butyl compound.

The effect of changes in bond order upon screening is exemplified by the fluorine resonances of the pair of diatomic molecules hydrogen fluoride, HF, and fluorine, F_2. The former possesses considerable ionicity in its bond while the latter is highly covalent. F_2 thus should have a large paramagnetic contribution to screening while HF should have little, and as expected HF is 630 ppm more highly screened than is F_2. Perversely, rather than being still more highly screened, the fully ionic fluoride ion, F, falls between F_2 and HF, but this probably reflects involvement of the fluoride ion in ion-solvent interactions so that it is not strictly a free ion. However in the case of gallium compounds the gallium resonance of the gallium (I) ion, Ga^+, does appear at higher applied magnetic field than any other gallium resonance so far reported. It is the most highly screened gallium nucleus and so has the least paramagnetic contribution to σ.

Usually the contributions to σ_d and σ_p for a nucleus are considered only for the electrons immediately neighbouring, or local to, that nucleus. More distant electrons give rise to long range effects on both σ_p and σ_d which are large but cancel to make only a small net contribution to σ. It is therefore more convenient to separate the long range effects into net contributions from different, quite localised, parts of the rest of the molecule. Two types of contribution to screening can be recognised and though they are small they are of particular importance for the proton resonance.

2.1.1 *Neighbour anisotropy effects*

We have already mentioned that in liquid samples due to the rapid and random motion of the molecules the magnetic fields at each nucleus due to all other magnetic dipoles average to zero. This is only true if the magnet (e.g. a nucleus) has the same dipole strength whatever the orientation of the

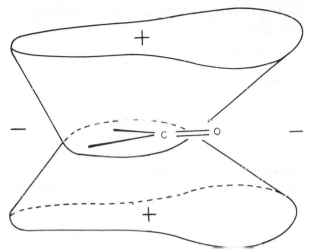

FIGURE 7

Screened and descreened volumes of space around a carbonyl bond. The sign + indicates that a nucleus in the space indicated would be more highly screened. The magnitude of the screening falls off with increasing distance from the group and is zero in the surface of the solid figure

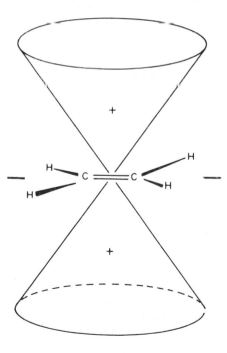

FIGURE 8

Screened and descreened volumes of space around a carbon-carbon double bond. The significance of the signs is the same as in Fig. 7. The cone axis is perpendicular to the plane containing the carbon and hydrogen atoms.

21

molecule relative to field direction. If the source of magnetism is anisotropic and the dipole strength varies with orientation in the applied field then a finite magnetic field appears at the nucleus.

Such anisotropic magnets are formed in the chemical bonds in the molecule, since the bonding electrons support different current circulation at different orientations of the bond axis to the field. The result is that nuclei in some parts of the space near a bond are descreened while in other parts the screening is increased. Figs. 7 and 8 show the way screening varies around some bonds.

A special case of anisotropic screening where the source of the anisotropy is clearly evident occurs in aromatic compounds and in acetylenes or nitriles, which exhibit what is called ring current anisotropy. The benzene structure for instance can support a large electronic ring current around the conjugated π bond system when the plane of the ring is transverse to the field axis but

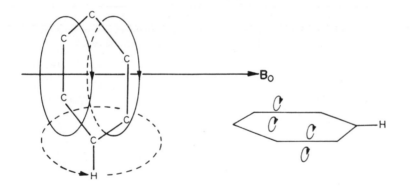

FIGURE 9

Ring current descreening in benzene. The area of the ring current loops due to the π-electrons is much smaller when the plane of the ring lies along the field axis and the reaction field is smaller. The reaction field is shown by a dashed line.

very little when the ring lies parallel to the field axis. This results in large average descreening of benzene protons since the average secondary magnetic field, which must oppose the applied field within the current loop, acts to increase the field outside the loop in the region of the benzene protons (Fig. 9). Similarly the two over-lapping π-bond systems of a triple bond can be regarded as supporting a ring current. The protons of acetylene however

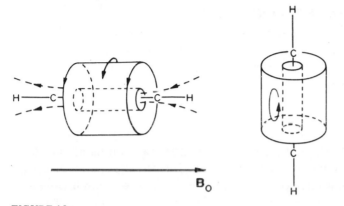

FIGURE 10

Ring current screening in acetylene where a large ring current is sustained around the C-C bond axis when this axis coincides with the applied field direction. The electron cloud is depicted as a cylinder.

in contrast to those of benzene are further screened by the ring current anisotropy since they lie on the ring current axis (Fig. 10).

We are now in a position to understand qualitatively the changes in screening constants $\Delta\sigma$ observed for protons in a number of hydrocarbons[*] (Fig. 11). The anisotropies, including that of the carbon-carbon single bond, play a considerable part in determining the screening constants.

The position of acetylene tends to emphasise the small effect of inductive electron drifts. The protons of acetylene are acidic and carry the least electron density yet they are highly screened due to the ring current anisotropy of the triple bond.

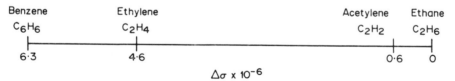

FIGURE 11

Changes in screening constants $\Delta\sigma$ for protons in some hydrocarbons with ethane arbitrarily taken as zero. Following convention the most highly screened protons are placed on the right so that increased $\Delta\sigma$ corresponds to reduced screening.

---

[*] The absolute value of σ is obtained only with difficulty.

2.1.2 *Through-space electric field effects*

Molecules which contain electric dipoles or point charges possess an electric field whose direction is fixed relative to the rest of the molecule. Such electric fields can perturb the molecular orbitals by causing electron drifts at the nuclei in the bond directions and by altering the electronic symmetry. It has been shown that the screening σ_E due to such electric fields is given by:

$$\sigma_E = - AE_z - BE^2 \qquad (2.3)$$

where A and B are constants, $A \gg B$, E_z is the electric field along a bond to the atom whose nuclear screening we require and E is the maximum electric field at the atom. The first term produces an increase in screening if the field causes an electron drift from the bond onto the atom and a decrease if the drift is away from the atom. The second term leads always to descreening. It is only important for proton screening in the solvation complexes of highly charged ions where E can be very large, though it is of greater importance for the nuclei of the heavier elements.

The electric field effect is of course attenuated with increasing distance. It is an intramolecular effect since the effect of external fields, for which the BE^2 term can be neglected, averages to zero as the molecule tumbles and E continually reverses direction along the bond.

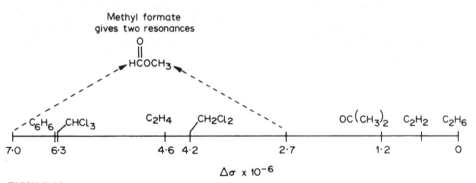

FIGURE 12

Changes in screening constants $\Delta\sigma$ for the protons in hydrocarbons and some molecules containing electric dipolar bonds. Note that increasing the number of carbon-chlorine bonds in the halomethanes results in a progressive decrease in σ. All but one of the compounds has been chosen with chemically indistinguishable protons and these give singlet resonances. Compounds with chemically distinguishable protons such as methyl formate give a resonance for each type of proton.

24

The descreening of protons which occurs in many organic compounds containing electronegative substituents X, probably occurs because of the electric field set up by the polar CX bond. This will also increase with X electronegativity and produce a similar result to an inductive electron drift.

Some typical changes in screening constants among such compounds are illustrated in Fig. 12 which is an extended version of Fig. 11. Notice how the screening is reduced as the distance between the hydrogen and the polar bond is decreased, HCC:O > HC.O or as the number of polar bonds is increased, CH_2Cl_2 > $CHCl_3$.

2.2 Effects due to unpaired electrons

The electron (spin = 1/2) has a very large magnetic moment and if, for instance, paramagnetic transition metal ions are present in the molecule large effects are observed. The n.m.r. signal of the nuclei present may be undetectable but under certain circumstances, when the lifetime of the individual electron in each spin state is short, so that its through space effect averages to near zero, n.m.r. spectra can be observed. The screening constants measured in such systems however cover a very much larger range than is normal for the nucleus and this arises because the electronic spins can be apparently delocalised throughout a molecule and appear at, or contact, nuclei. The large resonance displacements which result are known as contact shifts and the ligands in certain transition metal ion complexes exhibit proton contact shifts indicating several hundred ppm changes in σ. In addition, if the magnetic moment of the ion is anisotropic, one gets a through-space contribution to the contact interaction similar to the neighbour anisotropy effect and this is called a pseudo-contact shift. Study of contact effects has for some years been restricted to inorganic chemists but has recently been taken up by organic chemists also as a means of simplifying the proton spectra of organic molecules. An example is given on p 154.

2.3 The chemical shift

So far we have talked in terms of changes in screening of nuclei. It is however more usual to use the term 'chemical shift' rather than the cumbersome phrase 'the relative change in screening brought about by changes in chemical environment'. Thus if we have two nuclei in different environments with screening constants σ_1 and σ_2 then the two nuclear frequencies in a given magnetic field $\mathbf{B}_0$ are:

$$\nu_1 = \frac{\gamma \mathbf{B}_0}{2\pi} (1 - \sigma_1) \tag{2.4a}$$

$$\nu_2 = \frac{\gamma \mathbf{B}_0}{2\pi} (1 - \sigma_2) \tag{2.4b}$$

whence

$$\nu_1 - \nu_2 = \frac{\gamma \mathbf{B}_0}{2\pi} (\sigma_2 - \sigma_1) \tag{2.5}$$

We cannot measure the absolute value of $\mathbf{B}_0$ accurately enough for this relation to be of use, so we eliminate field from the equation by dividing through by ν_1 (equation 2.4a). This gives us the frequency change as a fraction of ν_1:

$$\frac{\nu_1 - \nu_2}{\nu_1} = \frac{\sigma_2 - \sigma_1}{1 - \sigma_1}$$

which since $\sigma_1 \ll 1$ reduces to:

$$\frac{\nu_1 - \nu_2}{\nu_1} = \sigma_2 - \sigma_1 \tag{2.6}$$

Thus the fractional frequency change is the same as the difference in screening in the two nuclear environments. This is called the chemical shift and is given the symbol δ. Its value is expressed in parts per million (ppm). It can be determined with high accuracy since it is possible to detect a frequency difference of 0·1 Hz in 100 MHz (0·001 ppm) for the narrow lines obtained with spin $\frac{1}{2}$ nuclei.

The chemical shift is calculated as follows: the frequency separation between the protons in benzene and methylene chloride is found to be 124·2 Hz at 60 MHz operating frequency ($\mathbf{B}_0 = 14\cdot1$ kilogauss (kG) or 1·41 T), then the chemical shift is:

$$\delta = \frac{124\cdot2}{60\,000\,000} \times 10^6 \text{ ppm} = 2\cdot07 \text{ ppm}$$

Or, more easily remembered, it is the shift in Hz divided by the operating frequency in MHz. The frequency separation will of course be different in a spectrometer operating at a different frequency, though δ will be the same

Because the chemical shift measurement is a relative term it is only possible to define a chemical shift scale for a nucleus if some substance is arbitrarily chosen as a standard and its δ set equal to zero. Thus in Fig. 12 ethane is the standard and the scale in units of $\Delta\sigma \times 10^6$ is also a ppm chemical shift (δ) scale.

The only nucleus for which a universally accepted standard has been chosen is the proton and the standard is tetramethylsilane, Me_4Si, usually abbreviated to TMS. This substance gives a narrow singlet resonance which does not interfere with most other proton resonances; it is miscible with most organic solvents; it is inert in most systems and, being highly volatile, can easily be removed after measurements have been made. Perversely to add to the difficulty of the subject, two chemical shift scales based on TMS are in common use, called the τ scale and the δ scale. Since most protons situated in organic compounds are less screened than TMS, the TMS signal is taken as origin in the δ scale and shifts are measured as being increasingly descreened. Thus δ for benzene is about 7·37 ppm. On the τ scale on the other hand TMS is taken as 10 ppm, the origin being an imaginary point some 2·63 ppm further descreened than benzene. Therefore benzene is at $\tau 2\cdot63$. Note that the units of τ are not usually quoted. The scales are related by $\tau = 10 - \delta$.

It is usual in recording spectra to depict the descreened region to the left-hand side of the spectrum with TMS to the right. The frequency then increases to the left of TMS since the magnetic fields at descreened nuclei are apparently higher and the nuclear precession frequency is higher. In older spectrometers with fixed frequencies and swept fields, the field had to be reduced to bring the precession frequency of these descreened nuclei into resonance. For this reason they were said to be low field, a term still very much in use, even for frequency swept spectra where they might more logically be called high frequency (Fig. 13). Because we have chosen our standard arbitrarily we also find that we have introduced sign into the δ and τ conventions. Thus in the τ scale the few resonances low field of $\tau = 0$ have negative shifts, while in the δ scale all those high field of TMS have negative shifts.

These δ and τ scales are only appropriate to proton spectra. For other

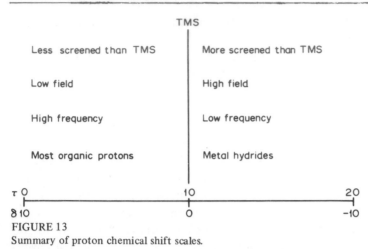

FIGURE 13

Summary of proton chemical shift scales.

nuclei except sometimes ^{31}P, ^{19}F, ^{13}C and ^{11}B the standard is often chosen as convenience dictates and the shifts recorded using the high field/low field terminology. 85 per cent orthophosphoric acid, H_3PO_4, is however, usually used as standard for ^{31}P resonances. It gives a broad line but has the advantage that it is easily available. Alternatively phosphorus oxide P_4O_6 which has a very narrow line may be used. In the case of ^{19}F the refrigerant $CFCl_3$ is commonly used either as standard or as both solvent and standard, and shifts referred to this substance are said to be on the ϕ scale. The ^{13}C resonance of TMS is often accepted as the standard for carbon spectroscopy, though as yet no symbol has been assigned to the scale using it as origin.* Boron spectra are standardised using primarily triemthyl borate, $(MeO)_3B$, or boron trifluoride etherate, $Et_2O.BF_3$.

The symbol δ is used extensively as short-hand for 'chemical shift' and when TMS is not the standard and its use in this way should not be confused with its use to designate the proton δ scale. Care should also be exercised in interpreting the meaning of any sign convention used, since examples are to be found in the literature in which + can mean either low field or high field. Usually it means the latter, the opposite convention to that of the proton δ scale. Currently it is suggested that we should adopt for all nuclei the convention that low field (i.e. high frequency) of standard is +.

* Acceptance now seems almost universal.

Internuclear spin–spin coupling 3

3.1 The mutual effects of nuclear magnets on resonance positions

The Brownian motion in liquid samples averages the through-space effect of nuclear magnets to zero. However, in the molecule $POCl_2F$ for example the phosphorus nucleus gives two resonances whose separation does not depend upon the magnetic field strength.[*] This suggests that the two resonances correspond to the two spin orientations of the fluorine nucleus and that the nuclei *are* able to sense one anothers magnetic fields. Theoretical considerations indicate that the interaction occurs via the bonding electrons. The contact between one nucleus and its s electrons perturbs the electronic orbitals around the atom and so carries information about the nuclear energy to other nearby nuclei in the molecule and perturbs their nuclear frequency. The effect is mutual and in the molecule above both the fluorine ($I = \frac{1}{2}$) and the phosphorus ($I = \frac{1}{2}$) resonances are split into doublets of equal Hz separation. The magnitude of the effect for a particular pair of nuclei depends on the following factors. (i) The nature of the bonding system, i.e. upon the number and bond order of the bonds intervening between the nuclei and upon the angles between the bonds. The interaction is not usually observed over more than five or six bonds and tends to be attenuated as the number of bonds increases though many cases are known where coupling over two bonds is less than coupling over three bonds. (ii) The magnetic moments of the two nuclei and is directly proportional to the product $\gamma_A \gamma_B$ where γ_A and γ_B are the magnetogyric ratios of the interacting nuclei. (iii) The valence s electron density at the nucleus and therefore upon the s character of the bonding orbitals. This factor also means that the interaction increases periodically as the atomic number of either or both nuclei is in-

[*] The chlorine nuclei ($I = 3/2$) have no effect. This is explained on p 85.

creased in the same way as does the chemical shift range.

The magnitude of the coupling interaction is measured in Hz since it is the same at all magnetic fields. It is called the coupling constant and is given the symbol J; its magnitude is very variable and values have been reported from 0·05 Hz up to several thousand Hz. The value of J gives information about the bonding system but this is obscured by the contribution of γ_A and γ_B to J. For this reason correlations between the bonding system and spin-spin coupling often use the reduced coupling constant, K, which is equal to $4\pi^2 J/(\hbar\gamma_A\gamma_B)$.

It is important to understand that coupling constants can be either positive or negative and that the frequency of one nucleus may be either increased or decreased by a particular orientation of a coupled nucleus, the sign depending upon the bonding system and upon the sign of the product $\gamma_A\gamma_B$.

Considerable data are available upon the magnitudes of interproton spin coupling constants from the mass of data accumulated for organic compounds. Interproton coupling is usually (though not always) largest between geminal protons, H.C.H., and depends upon the angle between the two carbon-hydrogen bonds. J_{gem} is typically 12 Hz in saturated systems. J falls rapidly as the number of intervening bonds is increased, being 7 to 8 Hz for vicinal protons (H.C.C.H.) and near zero across four or more single bonds. The same rules apply if oxygen or nitrogen forms part of the coupling path, and methoxy protons H_3C O CH R do not usually show resolvable coupling to the rest of the molecule though alcoholic or amino protons may do so to vicinal protons in e.g. HOCH. On the other hand coupling may be enhanced if there is an unsaturated bond in the coupling path, due to a σ-π configuration interaction and may be resolved over up to as many as nine bonds, e.g. ^9J(H-H) = 0·4 Hz between the hydrocarbon protons in $H_3C(C \equiv C)_3CH_2OH$. In saturated molecules a planar zig-zag configuration of the bonds may also lead to resolvable coupling over four or five single bonds. Note the use of the superscript 9 in the acetylene example to indicate the number of bonds over which the interaction occurs.

Karplus has calculated the values of the vicinal interproton coupling constants and shown that these depend upon the dihedral angle ϕ between the carbon-hydrogen bonds (Fig. 14). Two curves are shown to emphasize that the magnitude of coupling depends also upon the nature of the other substituents on the carbon, i.e. upon their electronegativity, their orientation, the hybridization at the carbon, upon the bond angles other than the

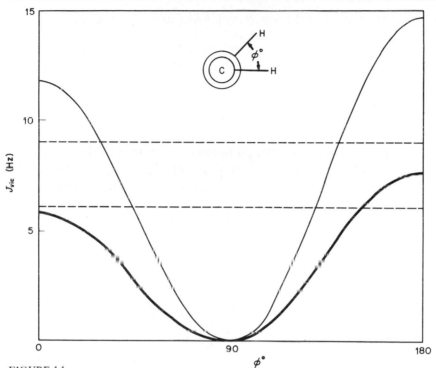

FIGURE 14

Karplus curves relating the dihedral angle ϕ in a HC-CH fragment and the vicinal proton-proton coupling constant. The inset diagram shows a view along the carbon-carbon bond. Two curves are shown relating to differently substituted fragments.and are differentiated by heaviness of line. The dotted lines show the typical range of values obtained when a group can rotate freely giving rise to an averaged J_{vic} (After Jackman and Sternhell.)

dihedral angle and upon the bond lengths. For instance, the vicinal coupling in ethyl derivatives decreases with increasing substituent electronegativity. J_{vic} is a rotational average of a Karplus curve because the methyl group rotates freely, and is 6·9 Hz in ethyl fluoride and 8·4 Hz in ethyl lithium. The curves cannot therefore be used to measure ϕ accurately but can often be used to distinguish a correct structure from a number of possibilities or to follow changes in conformation in closely related compounds. An example of this is given in Chapter 7, p 135 and the validity of the Karplus curve is demonstrated in Fig 60, p 108, though before we consider these we must find out how to recognize multiplicity which arises from spin coupling and how to determine J.

31

3.2 The appearance of multiplets arising from spin-spin coupling

The appearance of these multiplets is very characteristic and contains much information additional to that gained from the chemical shifts of each resonance. For this reason we intend to go into their analysis in some detail if not in depth. The simplest case to consider is the effect that a single chemically unique nucleus of $I = \frac{1}{2}$ has on other nuclei in the molecule that are sufficiently closely bonded. In half the molecules in the sample the spin of our nucleus N will be oriented in the same direction as the field and all the other nuclei in these molecules will have corresponding resonance positions. In the remainder of the molecules the spin of N will be opposed to the field and all the other nuclei in this half of the sample will resonate at slightly different frequencies to their fellows in the first half. Thus when observing the sample as a whole each of the nuclei coupled to N gives rise to two lines. The line intensities appear equal since the populations of N in its two states only differ by about 1 in 10^5 which is not detectable. We say that N splits the other resonances into 1 : 1 doublets. Because the z component of magnetization of N has the same magnitude in both spin states the lines are equally displaced from the chemical shift positions of each nucleus which are therefore at the centres of the doublets.

Let us in illustration consider the molecule $CHCl_2CH_2Cl$ and its proton resonance. This contains two sorts of hydrogen with the $CHCl_2$ proton resonating to low field of the CH_2Cl protons due to the greater electric field effect of two geminal chlorine-carbon bonds. The two CH_2Cl protons have the same frequency since rotation around the carbon-carbon bond averages their environments and makes them chemically and magnetically equivalent. We say they are isochronous. They are mutually spin coupled but because they are isochronous give rise to a singlet resonance in the absence of other coupling. We can state as a rule that isochronous protons always resonate as if they were a unit and give a singlet resonance unless coupled to other nuclei. This is why the substances depicted in Fig. 12 give rise to singlets despite the presence of coupling between the protons.

In the present example however the CH_2Cl resonance is split into a 1 : 1 doublet because of coupling to the non isochronous $CHCl_2$ proton. Equally, since the coupling interaction is mutual, the $CHCl_2$ proton is split by the two CH_2Cl protons, though the splitting pattern is more complex. We can discover

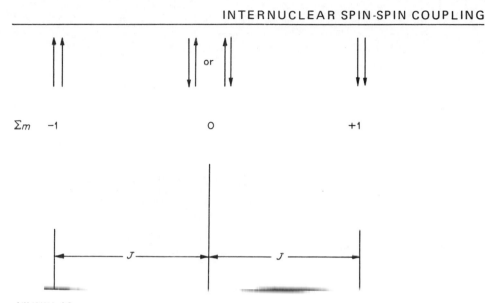

FIGURE 15

A stick diagram demonstrating the splitting due to two spin 1/2 nuclei. When $\Sigma m = 0$ there is no perturbation of the coupled resonance so that the centre line corresponds to the chemical shift position. This holds for all multiplets with an odd number of lines. The spacing J between the lines corresponds to a change in Σm of unity.

the shape of the $CHCl_2$ multiplet in several ways.

(i) By an arrow diagram (Fig. 15). In some molecules both the CH_2Cl spins will oppose the field, in others both spins will lie with the field while in the remainder they will be oriented in opposite directions. The $CHCl_2$ protons in the sample can each experience one of three different perturbations and their resonance will be split into a triplet. Since the CH_2Cl spins can be paired in opposition in two different ways there will be twice as many molecules with them in this state as there are with them in each of the other two. The $CHCl_2$ resonance will therefore appear as a 1 : 2 : 1 triplet with the spacing between the lines the same as that of the CH_2Cl 1 : 1 doublet. An actual spectrum is shown in Fig. 16 recorded on paper marked in the δ scale. Note that the line intensities are not exactly as predicted by this simple first-order theory due to more complex effects which we will discuss shortly. Note also that the total multiplet intensity is proportional to the number of protons giving rise to each multiplet.

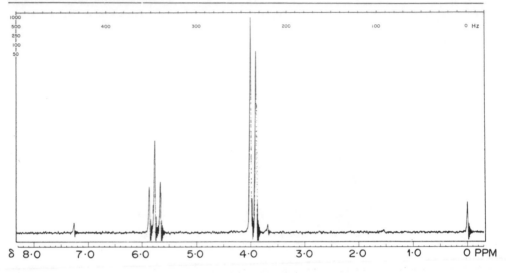

FIGURE 16

60 MHz proton spectrum of $CH_2ClCHCl_2$. The highest field resonance is TMS. The compound was dissolved in deuterochloroform (~ 7 per cent solution) and the $\frac{1}{2}$ per cent or so of protons in the solvent appears at 7·3 ppm.

(ii) We can also work out the multiplicity of the doublet by considering the possible values of the total magnetic quantum number Σm of the two CH_2Cl protons. $I = \frac{1}{2}$, and m can be $\pm \frac{1}{2}$, therefore for two protons (Fig. 15):

$$\Sigma m = +\tfrac{1}{2} + \tfrac{1}{2} = +1$$
$$\text{or } \Sigma m = +\tfrac{1}{2} - \tfrac{1}{2}$$
$$\text{or} \quad\quad -\tfrac{1}{2} + \tfrac{1}{2} \quad\Big\} = 0$$
$$\text{or } \Sigma m = -\tfrac{1}{2} - \tfrac{1}{2} = -1$$

Therefore we have three lines. Methods (i) and (ii) are equivalent, but method (ii) is particularly useful when considering multiplets due to muclei with $I > \frac{1}{2}$ where arrow diagrams become rather difficult to write down clearly. Note that when $\Sigma m = 0$ there is no perturbation of the chemical shift of the coupled group so that the centre of the spin multiplet corresponds to the chemical shift of the group.

Next let us consider the very commonly encountered pattern given by the ethyl group CH_3CH_2-. The isochronous pair of CH_2 protons are usually found low field of the CH_3 protons and are spin coupled to them. The CH_3

34

protons therefore resonate as a 1 : 2 : 1 triplet. The splitting of the CH_2X resonance caused by the CH_3 group can be found from Fig. 17, and is a 1 : 3 : 3 : 1 quartet. A typical ethyl group spectrum is shown in Fig. 18.

We have done enough now to formulate a simple rule for splitting due to groups of spin $\frac{1}{2}$ nuclei. Thus the number of lines due to coupling to n equivalent spin $\frac{1}{2}$ nuclei is $n + 1$. The intensities of the lines are given by the binomial coefficients of $(a + 1)^n$ or by Pascal's triangle which can be built up as required. This is shown in Fig. 19. A new line of the triangle is started by writing a 1 under and to the left of the 1 in the previous line and then continued by adding adjacent figures from the old line in pairs and writing

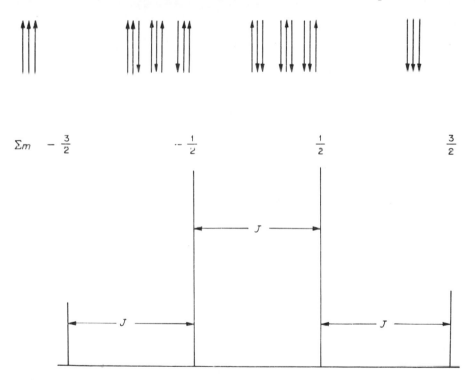

FIGURE 17

The splitting due to three spin 1/2 nuclei. There is no line which corresponds to $\Sigma m = 0$. The multiplet is however arranged symmetrically about the $\Sigma m = 0$ position so that the centre of the multiplet corresponds to the chemical shift position. This rule holds for all multiplets with an even number of lines

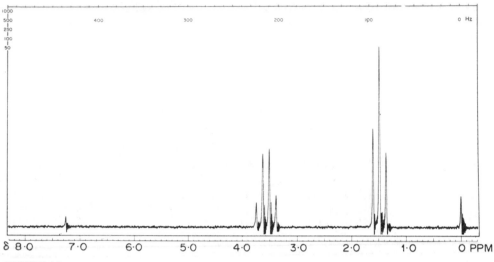

FIGURE 18
60 MHz proton spectrum of ethyl chloride CH_3CH_2Cl.

N								
0				1				Singlet
1			1		1			Doublet
2			1	2	1			Triplet
3		1	3	3	1			Quartet
4	1	4	6	4	1			Quintet

FIGURE 19
Pascals triangle can be used to estimate the intensities of the lines resulting from coupling to different numbers, N, of equivalent spin 1/2 nuclei. The figures in each line are obtained by adding adjacent pairs of figures in the line above.

down the sum as shown. The multiplicity enables us to count the number of spin $\frac{1}{2}$ nuclei in a group and the intensity rule enables us to check our assignment in complex cases where doubt may exist, since the outer com-

36

ponents of resonances coupled to large groups of nuclei (e.g. the CH of $(CH_3)_2CH\cdot$) may be too weak to observe.

Coupling to nuclei with $I > \frac{1}{2}$ leads to different relative intensities and multiplicities. In the case of a single nucleus the total number of spin states is equal to $2I + 1$ and this equals the multiplicity. If $I = \frac{1}{2}$ we get two lines, $I = 1$ gives three lines, $I = \frac{3}{2}$ gives four lines and so on. The spin populations of each state are virtually equal and so the lines are all of equal intensity and of equal spacing (Fig. 20 a-d).

Splitting due to multiple combinations of $I > \frac{1}{2}$ nuclei is much less common but a few examples have been recorded. Fig. 20e illustrates the calculation for two ^{11}B nuclei. The maximum total Σm is $\frac{3}{2} + \frac{3}{2} = 3$. $\Delta m = \pm 1$ and there are therefore 7 spin states so that the multiplicity of the signal from any coupled nuclei must be a septet. In order to determine the line intensities we have to find the number of ways each value of Σm can be obtained. This is shown in Table 3.1 and indicates relative intensities of $1 : 2 : 3 : 4 : 3 : 2 : 1$, which differs from the binomial distribution.

Line intensities for coupling to two nuclei with $I = \frac{3}{2}$

Σm	Possible spin combinations	Number of spin combinations
3	$+\frac{3}{2}+\frac{3}{2}$	1
2	$+\frac{1}{2}+\frac{3}{2}$ or $+\frac{3}{2}+\frac{1}{2}$	2
1	$+\frac{1}{2}+\frac{1}{2}$ or $+\frac{3}{2}-\frac{1}{2}$ or $-\frac{1}{2}+\frac{3}{2}$	3
0	$+\frac{3}{2}-\frac{3}{2}$ or $-\frac{3}{2}+\frac{3}{2}$ or $+\frac{1}{2}-\frac{1}{2}$ or $-\frac{1}{2}+\frac{1}{2}$	4
−1	$-\frac{1}{2}-\frac{1}{2}$ or $-\frac{3}{2}+\frac{1}{2}$ or $+\frac{1}{2}-\frac{3}{2}$	3
−2	$-\frac{1}{2}-\frac{3}{2}$ or $-\frac{3}{2}-\frac{1}{2}$	2
−3	$-\frac{3}{2}-\frac{3}{2}$	1

The rule given for spin $\frac{1}{2}$ nuclei can be generalized to include groups of nuclei of any I. The number of lines observed for coupling to n equivalent nuclei of spin I is $2nI + 1$.

More complex coupling situations also arise where a nucleus may be coupled simultaneously to chemically different groups of nuclei of the same or of different isotopes or species. The patterns are found by building up spectra, introducing the interactions with each group of nuclei one at a

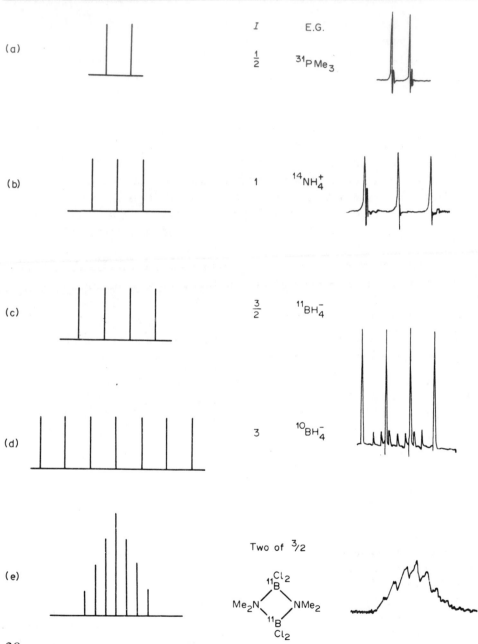

(a) I E.G.
$\frac{1}{2}$ $^{31}P\,Me_3$

(b) 1 $^{14}NH_4^+$

(c) $\frac{3}{2}$ $^{11}BH_4^-$

(d) 3 $^{10}BH_4^-$

(e)   Two of $^3/_2$

Me_2N — ^{11}B — Cl_2
Me_2N ⋯ NMe_2
^{11}B — Cl_2

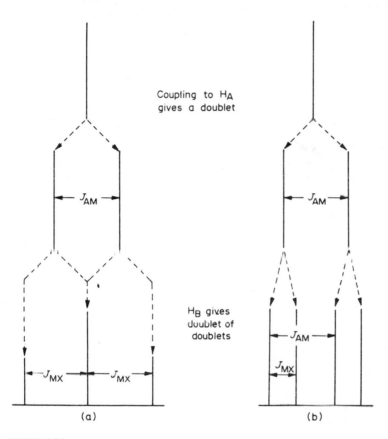

Coupling to H_A gives a doublet

J_{AM}

J_{AM}

H_B gives doublet of doublets

J_{AM}

J_{MX}

J_{MX}

J_{MX}

(a)

(b)

FIGURE 21

Splitting of the H_M protons of Z_2CH_A-$C(H_M)_2$-CH_XY_2 due to coupling to H_A and H_X: (a) $J(A-M) = J(X-M)$, the centre lines overlap and the multiplet is a 1 : 2 : 1 triplet just as if H_M were coupled to a CH_2 group; (b) $J(A-M) \neq J(A-M)$ and we get a doublet of doublets from which both $J(A-M)$ and $J(X-M)$ can be measured.

FIGURE 20

Multiplets observed in proton spectra due to coupling to nuclei with various spin quantum numbers. (a) – (d) single nuclei. Examples are given of the proton spectra of trimethylphosphine, of the ammonium ion and of the borohydride anion. Boron contains two isotopes ^{11}B ($I = \frac{3}{2}$) and ^{10}B ($I = 3$) and splitting due to both sorts of boron is observed. The coupling constant to ^{10}B is smaller than that to ^{11}B since the latter has a much larger magnetogyric ratio. (e) Shows the pattern for coupling equally to two ^{11}B nuclei. The ^{10}B multiplet in this case is not observable and leads only to line broadening. Coupling to nitrogen or chlorine is not observed.

39

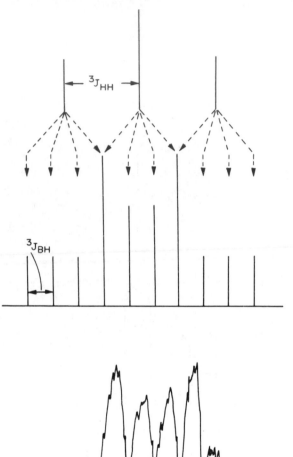

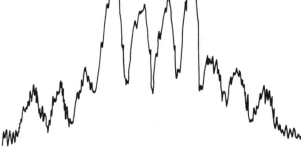

FIGURE 22

(a) Proton spectrum of the methyl group of $CH_3NH_2BF_3$. Overlap of lines occurs because $^3J(H\text{-}H) = 3 \times {^3J}(^{11}B\text{-}H)$. The superscripts to J refer to the number of bonds between the coupled nuclei. The methyl protons are coupled only to boron and the NH_2 protons. A typical trace is shown below. Coupling to ^{10}B leads to line broadening.

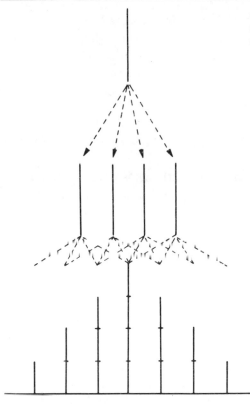

(b) An alternative way of predicting the splitting due to two equivalent spin $\frac{3}{2}$ nuclei as a quartet of overlapping quartets.

time. Thus Fig. 21 shows how a group M coupled to two chemically different spin $\frac{1}{2}$ nuclei A and X is split first by $J(A\text{-}M)$ into a doublet and shows that each doublet line is further split by $J(M\text{-}X)$. If $J(A\text{-}M) = J(A\text{-}X)$ a $1:2:1$ triplet is obtained, but if $J(A\text{-}M) \neq J(A\text{-}X)$ then a doublet of doublets with all lines of equal amplitude arises. This can be distinguished from a [11]B coupling because the line separations are irregular and of course the preparative chemist is usually aware whether or not there should be boron in his compound.

In analysing such multiplets it always has to be born in mind that overlap of lines may occur so that fewer than the theoretical number of lines are seen and the intensities are unusual. Such a case is illustrated in Fig. 22a for coupling to two equivalent protons and one boron nucleus [11]B. The boron gives a quartet splitting and each line of the quartet is split into a triplet by

41

the two equivalent protons so that twelve lines are expected. However because J(B-H) is fortuitously a sub-multiple of J(H-H) overlap occurs and only ten lines are observed. The same technique can be used to predict the shapes of multiplets due to several $I > \frac{1}{2}$ nuclei, introducing the effect of each nucleus one at a time. This has been done for two spin $\frac{3}{2}$ nuclei in Fig. 22b as an alternative to the Σm method.

3.3   Spin-spin coupling satellites

We would normally consider the spectrum of methane as being a singlet. In doing this we are of course thinking only of the molecule $^{12}C^1H_4$ and are ignoring the 1·1 per cent of $^{13}C^1H_4$ which contains the magnetically active ^{13}C isotope with $I = \frac{1}{2}$. If we looked carefully in the base line we would find a doublet with $J = 125$ Hz centred approximately on the intense singlet due to the main component. These two small lines are termed spin satellites and, while weak, can often be used to obtain extra information about otherwise symmetrical molecules. For example the molecule $(CF_3S)_3N$ has a singlet fluorine resonance with satellites due to molecules $^{13}CF_3SN(S^{12}CF_3)_2$. The fluorine in the $^{13}CF_3$ group is no longer isochronous with the $^{12}CF_3$ fluorine and the two sorts of fluorine atom can now exhibit spin coupling. As a result the satellites are split into septets by the other six fluorine nuclei bonded to ^{12}C (Fig. 23). This, together with the empirical formula, proves the presence of three chemically equivalent CF_3 groups. Note the magnitude of the fluorine-fluorine coupling constants which are often larger than proton-proton coupling constants over the same number of bonds.

The alteration in resonance position caused by the coupling interaction, in this case with ^{13}C, is often described as introducing an effective chemical shift since it results in the observation of spin coupling interactions which cannot be seen between the otherwise equivalent nuclei.

A second example is the spectrum of $W_2O_2F_9^-$ (Fig. 24). This consists of a doublet of intensity 8 and a nonet of intensity 1. The nine fluorine atoms can thus be divided into an isochronous set of eight and one unique atom. There are in addition however spin satellite lines due to coupling to the 14·28 per cent of ^{183}W which has a spin $I = \frac{1}{2}$. These lines will originate from the ions $^{183}W^{184}WO_2F_9^-$ since only 2 per cent of the molecules will

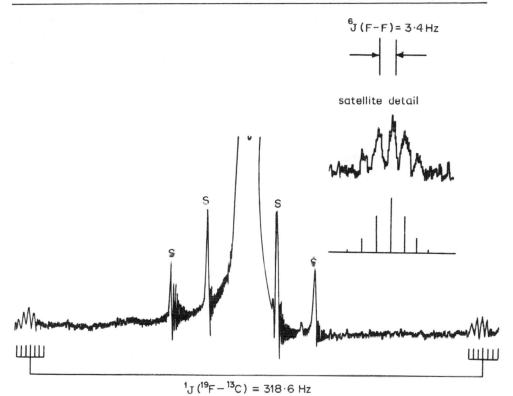

FIGURE 23

The ^{19}F spectrum of $(CF_3S)_3N$ showing the satellites due to 1·1 per cent of $(^{13}CF_3S)N(^{12}CF_3S)_2$. The fluorine on ^{13}C is split into a septet by long range coupling with the fluorine on ^{12}C, $^6J(F-F) = 3·4$ Hz, while these latter are mutually split into a quartet. This is lost beneath the singlet due to the all ^{12}C molecule. The lines marked S are spinning sidebands.

contain two ^{183}W atoms and their resonance will be much weaker. Each of the lines of the intense doublet has two ^{183}W satellites, each of which is further split into a 1 : 4 : 6 : 4 : 1 quintet. This pattern must arise from coupling to four fluorine atoms. We can therefore conclude that we have four of the eight isochronous fluorine atoms associated with the ^{183}W atom and therefore split into a satellite doublet and then further coupled to the remaining four which are equally associated with the ^{184}W atom. This provides considerable confirmatory evidence that the structure is $OWF_4 \cdot F \cdot WF_4O$ with a fluorine atom bridging the tungsten atoms.

43

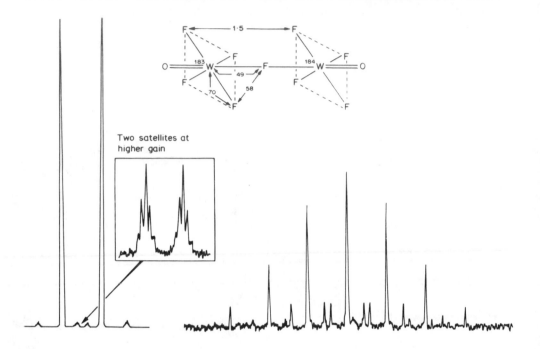

FIGURE 24

[19]F spectrum of $W_2O_2F_9{}^-$ showing spin satellites due to 14·28 per cent of [183]W. The outer lines of the nonet are lost in the base-line noise and the student should confirm that the intensity ratios of the observed lines correspond to those expected for the inner seven of the nonet rather than to those expected for a septet. The arrows around the formula indicate the various coupling interactions in Hz. The single fluorine nonet is recorded at higher gain. (After McFarlane, Noble, and Winfield.)

3.4 Second-order effects

The rules so far discussed apply to spectra of nuclei of the same species where the chemical shifts or the effective chemical shifts are large compared with the coupling constants, or to coupling between nuclei of different species where the frequency separation of lines is invariably large. A nomenclature has been adopted for these cases in which the chemically non-equivalent sets of spins are labelled with letters from the alphabet, choosing letters that are well separated in the alphabetic sequence to signify large chemical shift separation. Thus $CH_2ClCHCl_2$ is an A_2X system, CH_3CH_2R is an A_3X_2 system, and CH_3CH_2F is an A_3M_2X system. Their spectra are called first order.

We have already seen in the examples that the line intensities in proton spectra exibiting interproton coupling often do not correspond exactly to those predicted by the first order rules and these distortions increase as the interproton chemical shift is reduced. The spectra are said to become second order and to signify this and the fact that the chemical shifts between the coupled nuclei are relatively small, the spins are labelled with letters close together in the alphabet. Thus for example, two coupled protons resonating close together are given the letters AB and an ethyl group in $(CH_3CH_2)_3Ga$, where the methyl and methylene protons resonate close together, is described as an A_3B_2 grouping. Mixed systems are also possible and a commonly encountered one is the three spin ABX grouping where two nuclei resonate close together and a third is well shifted or is of a different nuclear species.

Second-order spectra arise when the frequency separation between multiplets due to different equivalent sets of nuclei is similar in magnitude to the coupling constant between them; under these circumstances the effects due to spin coupling and chemical shift have similar energy and become intermingled, leading to alterations in relative line intensities and in line positions. Because it is the ratio between the frequency separation and J which is important, chemical shifts are always expressed in Hz and not in ppm. The Hz separation is obtained by multiplying the chemical shift, δ, by the spectometer operating frequency and is written $\nu_0\delta$. The perturbation of the spectra from the first-order appearance is then a function of the ratio $\nu_0\delta/J$ and is different for spectrometers operating at different frequencies. If a high enough frequency is used, many second-order spectra approach their first-order arrangement.

45

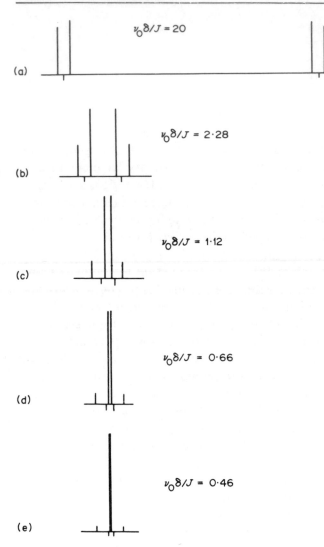

FIGURE 25

AB quartets for several values of the ratio $\nu_0 \delta/J$. The small markers under the base-line represent the true A and B chemical shift.

We will consider the simplest possible system with two spin $\frac{1}{2}$ nuclei, i.e the AB system. When the chemical shift between them is large then we see two doublets. A typical arrangement with $J = 10\,\text{Hz}$ and $\nu_0\delta = 200\,\text{Hz}$ is shown in Fig. 25a. If we reduce the chemical shift progressively to zero we can imagine the two doublets approaching one another until they coincide. However we know that an isolated pair of equivalent nuclei give rise to a singlet and the problem is how can two doublets collapse to give a singlet. If the coupling constant remains at $10\,\text{Hz}$ why do we not get a doublet A_2 spectrum?

The behaviour of the multiplets can be predicted using a quantum mechanical argument. The system can have any one of four energy states. These are characterized by the spin orientation of the two nuclei. The wave functions of the two spin states are normally written α for $I = +\frac{1}{2}$ and β for $I = -\frac{1}{2}$. There are four such spin states with the wave functions $\alpha\alpha$, $\alpha\beta$, $\beta\alpha$ and $\beta\beta$, where the first symbol in each pair refers to the state of the A nucleus and the second symbol to the state of the B nucleus. The energies of the states $\alpha\alpha$ and $\beta\beta$ can be calculated straight forwardly but the two states $\alpha\beta$ and $\beta\alpha$ have the same total spin angular momentum and it is found that the quantum mechanical equations can only be solved for two linear combinations of $\alpha\beta$ with $\beta\alpha$. This is described as a mixing of the states and means that none of the observed transitions corresponds to a pure A or a pure B transition. The form of the wave functions and the energy levels derived are shown below. C_1 and C_2 are constants, ν_A and ν_B are the A and B nuclear frequencies in the absence of coupling and $\nu_0\delta = |\nu_A - \nu_B|$.

No.	Wave function	Energy level (Hz)
1	$\alpha\alpha$	$\frac{1}{2}(\nu_A + \nu_B) + \frac{1}{4}J$
2	$C_1\,(\alpha\beta) + C_2\,(\beta\alpha)$	$\frac{1}{2}[(\nu_0\delta)^2 + J^2]^{1/2} - \frac{1}{4}J$
3	$-C_2\,(\alpha\beta) + C_1\,(\beta\alpha)$	$-\frac{1}{2}[(\nu_0\delta)^2 + J^2]^{1/2} - \frac{1}{4}J$
4	$\beta\beta$	$-\frac{1}{2}(\nu_A + \nu_B) + \frac{1}{4}J$

The transition energies are the differences between four pairs of energy states, 3-4, 2-4, 1-3, and 1-2, each transition involving a mixed energy level. Because of the mixing the transition probabilities are no longer equal as in the first-order case and intensity is transferred from lines in the outer parts of the total multiplet into the central region. The transition energies relative to the centre of the multiplet, i.e. to the mean frequency $\frac{1}{2}(\nu_A + \nu_B)$, and the intensities are:

	Transition	Energy(Hz)	Relative intensity
a	3→1	$+\frac{1}{2}J + \frac{1}{2}[(\nu_0\delta)^2 + J^2]^{1/2}$	$1 - J/[(\nu_0\delta)^2 + J^2]^{1/2}$
b	4→2	$-\frac{1}{2}J + \frac{1}{2}[(\nu_0\delta)^2 + J^2]^{1/2}$	$1 + J/[(\nu_0\delta)^2 + J^2]^{1/2}$
c	2→1	$+\frac{1}{2}J - \frac{1}{2}[(\nu_0\delta)^2 + J^2]^{1/2}$	$1 + J/[(\nu_0\delta)^2 + J^2]^{1/2}$
d	4→3	$-\frac{1}{2}J - \frac{1}{2}[(\nu_0\delta)^2 + J^2]^{1/2}$	$1 - J/[(\nu_0\delta)^2 + J^2]^{1/2}$

There are thus four lines as in the AX spectrum but with perturbed intensities. The line positions and the corresponding energy level diagram are shown in Fig. 26. The energy levels are marked with the appropriate spin state and the transitions which in the first-order case can be regarded as arising from transitions of the A or B nucleus form opposite sides of the figure. The three line separations are $a\text{-}b = J$, $c\text{-}d = J$, and $b\text{-}c = [(\nu_0\delta)^2 + J^2]^{\frac{1}{2}} - |J|$. The separation $a\text{-}c$ or $b\text{-}d$ which in the first-order case is the same as the separation between the doublet centres, and is therefore the chemical shift in Hz, $\nu_0\delta$, is now simply $[(\nu_0\delta)^2 + J^2]^{\frac{1}{2}}$ and is larger than the true chemical shift. In other words though $\nu_0\delta$ is reduced to zero the doublet centres never coincide and are separated by J Hz. The outer lines however have intensity zero at this point while the inner lines are coincident, i.e. we predict a singlet spectrum as is observed. C.f. Fig 25e. Thus arises our rule 'isochronous coupled protons resonate as a unit'.

We can calculate some simple rules for analysing an AB spectrum.

(i) The spectrum contains two intervals equal to J. $a\text{-}b$ and $c\text{-}d$.

(ii) The true AB chemical shift $\nu_0\delta$ is found as follows:

$$
\begin{aligned}
(a\text{-}d)\,(b\text{-}c) &= ([(\nu_0\delta)^2 + J^2]^{\frac{1}{2}} + J)([(\nu_0\delta)^2 + J^2]^{\frac{1}{2}} - J) \\
&= (\nu_0\delta)^2 + J^2 - J^2 \\
&= (\nu_0\delta)^2 \\
\therefore \quad \nu_0\delta &= [(a\text{-}d)\,(b\text{-}c)]^{\frac{1}{2}}
\end{aligned}
$$

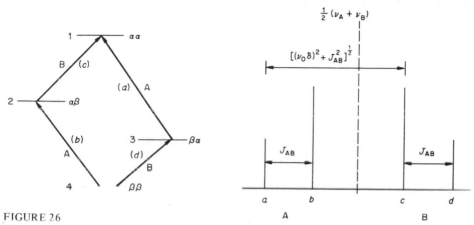

FIGURE 26

The energy level diagram for a system of two spins related to the resulting AB quartet. The spectrum was calculated for ν 10·0 Hz and δ 20·2 Hz, i.e. $\nu_0 \delta$ 2·02.

a-d is the separation between the outermost lines and *b-c* is the separation between the innermost pair of lines.

(iii) The assignment can be checked against the intensity ratios of the larger and smaller lines. The intensity ratio, stronger/weaker is:

which gives:

$$\left(1 + J/[(\nu_0\delta)^2 + J^2]^{\frac{1}{2}}\right) / \left(1 - J/[(\nu_0\delta)^2 + J^2]^{\frac{1}{2}}\right)$$

$$\left([(\nu_0\delta)^2 + J^2]^{\frac{1}{2}} + J\right) / \left([(\nu_0\delta)^2 + J^2]^{\frac{1}{2}} - J\right)$$

the ratio of the line separations $(a-d)/(b-c)$.

Note that changing the sign of J does not alter the pattern.

Fig. 25a-e shows the form of the AB quartet for several values of $\nu_0\delta/J$. The true chemical shift positions are marked below the base-line. It is important to remember that if a multiplet shows signs of being highly second order then both intensities *and* resonance positions are perturbed from their first-order values. A spacing corresponding to J_{AB} remains in AB type spectra since only one coupling interaction exists, but in more complex systems the spacings are combinations of coupling constants. On the other hand if the intensity perturbation is only slight (Fig. 25a) then the line positions are not detectably perturbed. Three examples of actual spectra which contain an AB multiplet are given in Fig. 27.

49

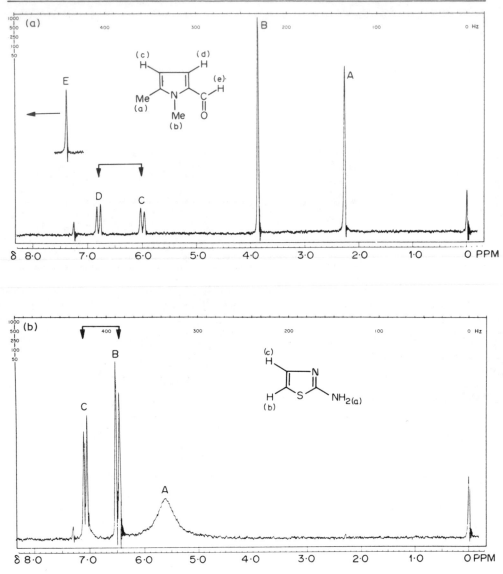

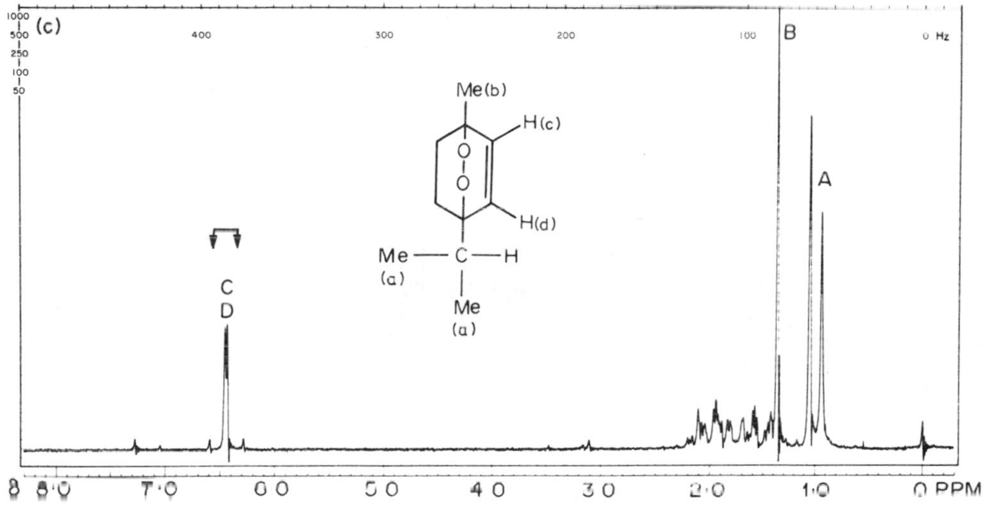

FIGURE 27

60 MHz spectra of some compounds containing AB groupings of protons, in order of decreasing $v_0 \delta / J$. The AB quartets are at low field in each case and are bracketed. (a) 1, 5-dimethyl pyrrole-2-aldehyde. E is offset and is observed at 9·38 ppm. Protons (C) and (D) give the AB quartet. (b) 2-aminothiazole. Note the broad line arising from hydrogen on ^{14}N. Protons (B) and (C) give the AB quartet. (c) Ascaridole. Protons (C) and (D) give the AB quartet.

3.4.1 The three spin system

We shall first construct an energy level diagram for the three spin system similar to that of Fig. 26 for the AB system (Fig. 28). The spin orientations are indicated by the signs + or − and are set down in groups of all the possible combinations of three spins, with groups of different total angular momentum placed on different levels. The spins are always written in the order spin 1, spin 2, spin 3 and as laid out the transitions + → − of any one spin form the four parallel sides of a cube. Thus the transitions A_1 A_2 A_3 and A_4 occur for a + → − transition of spin 1 in the presence of the spin 2, spin 3 combinations $++, -+, +-, --$.

The cube diagram of Fig. 28 is used extensively in analysing three spin spectra. We shall use it to demonstrate how degeneracy arises in n.m.r. spectra and to indicate how the system ABX may be analysed.

51

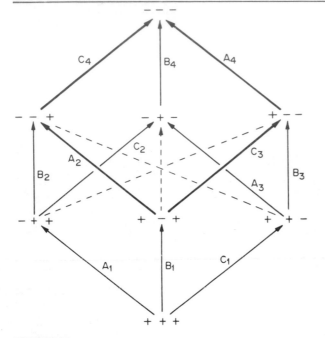

FIGURE 28

An energy level diagram for a three spin ($I = \frac{1}{2}$) system. Each edge of the cube corresponds to one of the twelve possible transitions. The dashed lines indicate combination transitions where all three spins change orientation simultaneously though Δm remains unity.

First we shall consider how we can obtain the form of the first-order AX_2 spectrum from the cube. If we let spin 1 be nucleus A and spins 2 and 3 be the nuclei X_2 then spins 2 and 3 are isochronous and equally coupled to spin 1. We obtain transitions as follows:

One A transition for X_2 oriented $+ +$ (line A_1)
Two A transition for X_2 oriented $+ -$ (line A_2 and
          or $- +$ A_3)
(These are now indistinguishable)
One A transition for X_2 oriented $- -$ (line A_4)
      giving a 1 : 2 : 1 triplet

Two X_2 transitions for A oriented $+$ (lines $B_1C_1B_2C_2$)
Two X_2 transitions for A oriented $-$ (lines $B_3C_3B_4C_4$)
      giving a 4 : 4 doublet

We see two X_2 transitions rather than four X transitions since the two nuclei resonate as a unit.

The above description of the spectrum shows that the centre A transition and the two X_2 transitions each consist of two coincident transitions, i.e. they are degenerate transitions. This enables us to understand qualitatively why, when we reduce the A-X chemical shift so as to give a second-order

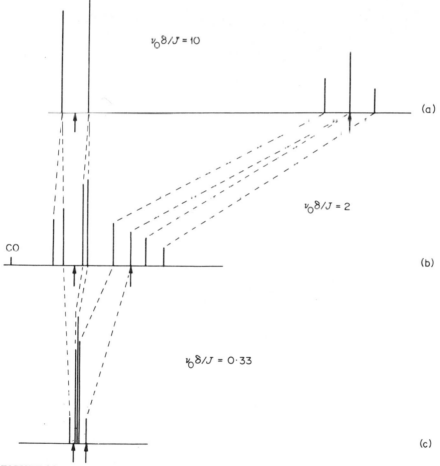

FIGURE 29

Diagram showing how the degenerate lines in an AB_2 spectrum separate as the chemical shift is reduced. In c the flanking lines are too weak to be observed. CO is a combination line.

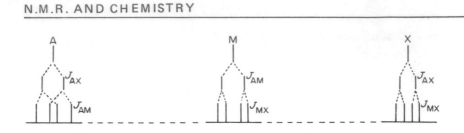

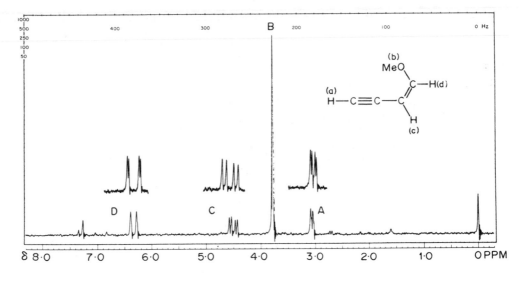

FIGURE 30

The spectral first-order pattern expected for coupling between three nuclei with three different coupling constants (1, 2, and 3 units in this case). An example of an AMX spectrum in 1-methoxy-1-buten-3-yne is also shown. The upper multiplets have been re-run on an expanded scale.

AB_2 spectrum in which the line positions are perturbed, we observe an eight line spectrum. The degeneracies are lifted and we can resolve four A transitions and four B_2 transitions. The line positions and intensities calculated for different $v_0\delta/J$ are given in Fig. 29. A ninth line is depicted for $v_0\delta/J = 2$. This is a combination line which corresponds to a combined transition of all three spins such that $\Delta\Sigma m = 1$, c.g. $- + + \rightarrow + - -$. There are three such possible transitions marked by dashed arrows in Fig. 28, which may be observed in second-order spectra. An AB_2-like spectrum is shown in Fig. 77.

If we make our three nuclei have different chemical shifts with three different coupling constants then in the first-order case, AMX, we have 12 lines. These are represented by the 12 sides of the cube diagram. None of the transitions coincide since none of the $+-$ pair combinations are equivalent (Fig. 30).

If we allow the M nucleus to have a similar frequency to the A nucleus ⬛⬛⬛⬛⬛⬛⬛⬛⬛⬛⬛ AB ⬛⬛⬛⬛⬛⬛ The AB ⬛⬛⬛⬛⬛⬛⬛⬛⬛⬛ The close AB coupling leads to second-order perturbation in the X region of the spectrum also. The analysis of the AB part can be accomplished fairly easily if we consider it to be made up of two AB sub-spectra i.e. one quartet corresponding to X spin $= +\frac{1}{2}$ and one corresponding to X spin $= -\frac{1}{2}$. This is equivalent to separating the cube diagram into two AB diagrams connected by the C transitions. The effective chemical shifts E and E^* used to calculate the line positions and intensities of the AB sub-spectra are, for one (E) $v_0\delta_A - \frac{1}{2}J_{AX}$ and $v_0\delta_B - \frac{1}{2}J_{BX}$ and for the other (E^*) $v_0\delta_A + \frac{1}{2}J_{AX}$ and $v_0\delta_B + \frac{1}{2}J_{BX}$. Since $J_{AX} \neq J_{BX}$ the frequency separation of the A and B resonances are not the same in the two AB sub-spectra, the value of $[(v_0\delta)^2 + J_{AB}^2]^{\frac{1}{2}}$ is not the same and no line separations corresponding to $J(A-X)$ or $J(B-X)$ can be found in the AB part. This is demonstrated in the scale diagram of Fig. 31. Four separations $J(A-B)$ can however be recognized.

The X part, which can arise from the same or a different nuclear species as A and B, consists of four transitions plus two combination lines. Again no spacing corresponding to $J(A-X)$ or $J(B-X)$ exists in the X part though one can be found which corresponds to their sum. The spectrum can be analysed from this and from the effective chemical shifts obtained from the AB part.

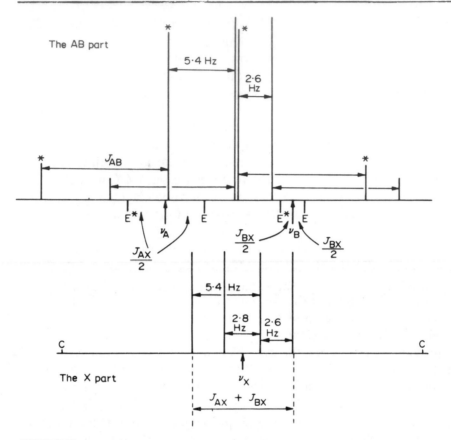

FIGURE 31

Diagram of an ABX spectrum. One AB sub-spectrum is differentiated from the other by the asterisk. The parameters used to calculate line positions and intensities are $|\nu_A - \nu_B| = 10$ Hz, $J(A-B) = 10$ Hz, $J(A-X) = 6$ Hz, and $J(B-X) = 2$ Hz. The positions E, E^* give the effective chemical shifts of each of the AB sub-spectra and are shifted from the true resonance positions ν_A, ν_B by amounts depending on $J(A-X)$ and $J(B-X)$. Because none of E or E^* coincides with the centres of the AB doublets, none of the spacings is equal to $J(A-X)$ (6 Hz) or $J(B-X)$ (2 Hz). Four spacings equal to $J(A-B)$ can however be observed.

It should be noted that if the sign of $J(B-X)$ is different from that of $J(A-X)$ the line order is altered and the form of the spectrum changes quite drastically. We can thus obtain relative signs of coupling constants in the ABX case.

3.4.2. *Deceptively simple spectra*

Sometimes a set of nuclei give rise to a second-order spectrum which because of special values of the shifts and coupling constants looks very like a first-order spectrum. These are called deceptively simple spectra, and if interpreted on a first-order basis could give the wrong parameters. Two cases are given in illustration based on the ABX case.

(*i*) If either $\nu_0\delta_{AB} = 0$ or $J_{AX} = J_{BX}$

Before proceeding further we have to extend our definition of equivalent nuclei. If $\nu_0\delta_{AB} - 0$ then the AB nuclei are isochronous and we might expect an A_2X spectrum. However the two nuclei are not fully magnetically equivalent because $J(A\text{-}X) \neq J(B\text{-}X)$ and they are each coupled differently to the X nucleus. For full magnetic equivalence nuclei must be isochronous *and* coupled equally to each of the other non-isochronous nuclei of the set.

Returning to our ABX case we find that either of the above situations means that we get a symmetrical AB part and that in the X region the two central lines overlap giving a $1:2:1$ triplet. The observation of the X triplet need not mean that $J(A\text{-}X) = J(B\text{-}X)$ while the AB spectrum may persist despite the fact that A and B are isochronous.

(*ii*) If $\nu_0\delta_{AB}$ and $\frac{1}{2}(J(A\text{-}X) - J(B\text{-}X))/J(A\text{-}B)$ are both small then we also get an X triplet and the outer lines of the AB sub-spectra become too small to observe thus giving the appearance of an A_2X spectrum, though the spacings do not correspond to any coupling constant.

3.4.3 *Virtual coupling*

The typical AMX spectrum consists of twelve lines of equal intensity. Because the intensities are equal we know it is a first-order spectrum and that the coupling constants can be measured directly from the line separations. Now let us see in more detail what happens as we allow the M resonance to approach the A resonance to give an ABX spectrum (Fig. 32). The perturbation of the X part of the spectrum which occurs as the AM nuclei approach and become AB can be clearly seen in the figure. We say that when $J(A\text{-}B)/\nu_0\delta$ is large the system is strongly coupled.

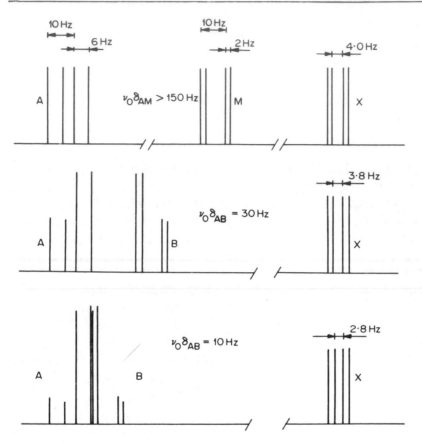

FIGURE 32

Changes which occur in the spectrum of three spin $\frac{1}{2}$ nuclei from pure first-order AMX to second-order ABX. The coupling parameters are the same as in Fig. 31.

Now let us consider what would happen to the X spectrum if $J(B\text{-}X) = 0$. In the weakly coupled AMX case with $J(M\text{-}X) = 0$ the M and X resonances are split into doublets giving a total of eight lines (Fig. 33a). However if the AM part becomes strongly coupled (AB) all twelve lines of an ABX spectrum appear (Fig. 33b). This is often described as virtual coupling, i.e. X and B are apparently coupled via the strong AB interaction. This often leads to increased complexity in spectra which otherwise might be relatively

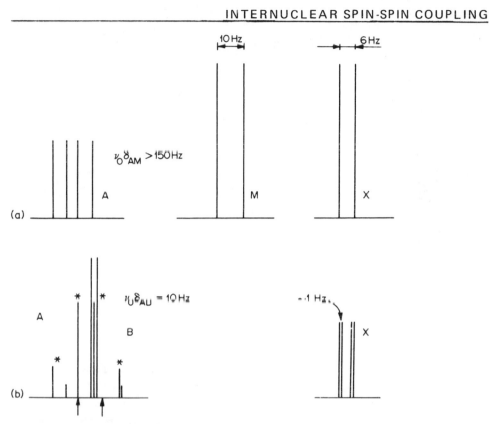

FIGURE 33

Spectrum of a three spin system with one coupling constant zero. (a) First order with $J(\text{M-X}) = 0$. (b) In the second-order ABX case with $J(\text{B-X}) \doteq 0$, second-order perturbation still occurs in the X part to give four lines. This is described as virtual coupling via the strong AB interaction. The parameters used are $J(\text{A-B}) = 10$ Hz, $J(\text{A-X}) = 6$ Hz, $J(\text{B-X}) = J(\text{M-X}) = 0$ Hz, and $|\nu_A - \nu_B| = 10$ Hz.

simple. For instance the methylene protons of long-chain hydrocarbons often appear as broad bands of overlapping lines rather than as a series of near coincident quintets as might at first be expected.

3.4.4. Spin-spin satellites and second-order effects

Spin satellites may show second-order splitting. For instance dioxane contains two sets of four isochronous protons (Fig. 34). A ^{13}C atom at one position splits the resonance of its attached A protons into a doublet. If we consider

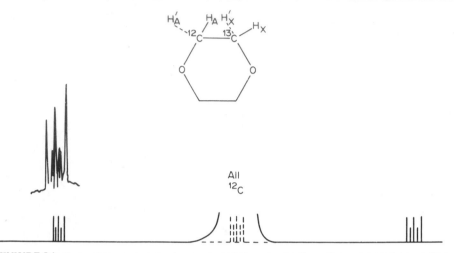

FIGURE 34

The ^{13}C satellite spectrum of dioxane. The ring structure prevents complete rotation of the methylene groups so that $J(A-X)$ and $J(A'-X)$ cannot average to the same value. Protons A and A' are thus not fully equivalent. The inset shows an actual satellite trace. (After Sheppard and Turner.)

the molecules with the ^{13}C nucleus all in one orientation, i.e. those giving rise to one satellite, then these contain two isochronous A nuclei coupled to two isochronous X nuclei resonating at a different frequency. The nucleus A is coupled to X and to X' by different amounts since free rotation is not possible, so that X and X' though isochronous are not magnetically equivalent. Similarly A and A' are not equivalent. The spectrum is therefore second-order even though the AX effective chemical shift is quite large.

Note the use of primes in the above example to differentiate isochronous but magnetically non-equivalent nuclei. In large spin systems the use of primes can become cumbersome and so an alternative nomenclature has been recently introduced. This would describe the above system as an $[AX]_2$ system and examples using these square brackets will be met more often in the future.

A more complex set of second-order spin satellites is observed in the spectrum of benzene and are illustrated here to demonstrate the degree of complexity that can arise. They are usually observed as low intensity, broad resonances and originate from a six spin XAA'BB'C system. The

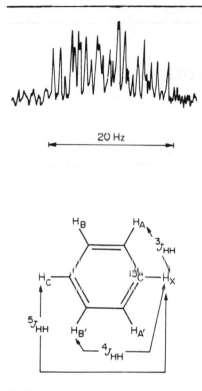

FIGURE 35

A ^{13}C satellite of benzene which illustrates the complexity which may arise in second-order spectra of several spins.

satellites, each of which theoretically contains 45 lines, Fig. 35, have been resolved and values calculated for all the proton-proton and certain combinations of the carbon-proton coupling constants:

^3J(H-H) $=$ 7·54 Hz
^4J(H-H) $=$ 1·37 Hz
^5J(H-H) $=$ 0·69 Hz
^1J(^{13}C-H) $=$ 158·34 Hz
^3J(^{13}C-H) $-$ ^2J(^{13}C-H) $=$ 6·36 Hz
^4J(^{13}C-H) $-$ ^2J(^{13}C-H) $=$ $-$ 2·21 Hz

61

Nuclear magnetic relaxation and related phenomena

4

4.1　Relaxation processes in assemblies of nuclear spins

If we perturb a physical system from its equilibrium condition and then remove the perturbing influence, the system will return to its original equilibrium condition. It does not return instantaneously however but takes a finite time to readjust to the changed conditions. The system is said to relax. Relaxation to equilibrium usually occurs exponentially, following a law of the form:

$$(n - n^e)_t = (n - n^e)_0 \exp(-t/T)$$

where $(n - n^e)_t$ is the displacement from the equilibrium value n^e at time t and $(n - n^e)_0$ that at time zero. The relaxation rate can be characterised by a characteristic time T. If T is small, relaxation is fast, whereas if T is is large, relaxation is slow.

The concept of the relaxation times appropriate to assemblies of magnetically active nuclei is of high importance to our technique and allows us to understand a considerable number of n.m.r. phenomena. For this reason we intend to cover the subject with some care. We have already noted that in an assembly of spin $\frac{1}{2}$ nuclei the spins in the two energy levels are in equilibrium, a small excess number existing in the low energy level. In Fig. 36a these nuclei are shown precessing around a conical surface, and evenly distributed over it. This constitutes an equilibrium condition for our system, with net magnetization M_z in the longitudinal or field axis direction and

FIGURE 36

A system of nuclear magnets ($I = \frac{1}{2}$) (a) can be perturbed in two ways, either by altering the field B_0(c) when the excess population in the lower energy state changes at a rate determined by T_1, or by tipping the precession cone off axis (b) when a return to equilibrium is governed by T_2.

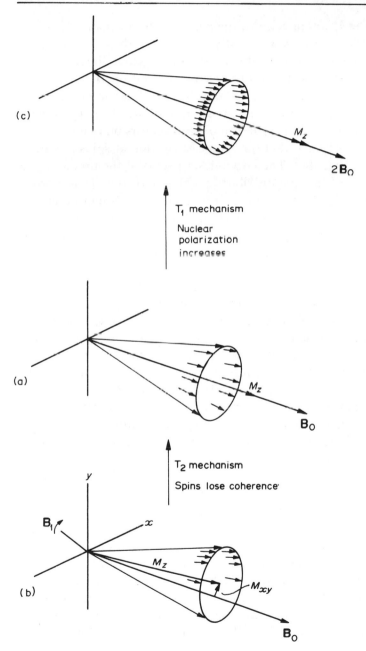

zero net magnetization M_{xy} transverse to the z axis in the xy plane. This system can be perturbed in two ways. Firstly we can introduce a rotating field B_1 rotating at the same angular velocity and in the same direction as the nuclei, which tips the cone axis to produce magnetization M_{xy}. This is the mechanism described previously to account for the detection of a nuclear signal. An alternative view of the production of M_{xy} is that the precession around B_1 results in a concentration of spin vectors on one side of the Larmor precession cone and that the even distribution of spins around the cone is perturbed (Fig. 36b). These two descriptions of the process are equivalent but the idea of the spin distribution being perturbed is of more use when considering relaxation processes. If B_1 is small the longitudinal magnetization is hardly affected. If B_1 is switched off or moved out of resonance the system relaxes to equilibrium with $M_{xy} = 0$ (Fig. 36a) and the spins become evenly distributed around the cone surface again. This takes time since it involves the spins moving with different angular frequencies. This process, affecting the transverse magnetization is known as transverse relaxation and its characteristic time is given the symbol T_2.

Secondly we can consider what might happen if we were suddenly to double B_0. This would increase the energy separation between the two spin states and so increase the number of spins expected to remain in the low energy state. In other words the new equilibrium value of longitudinal magnetization M_z is increased. The system will relax to the new value by means of spins undergoing transitions from the upper to the lower energy level. This involves a loss of energy by the system which also requires a finite time. This process, which affects only the longitudinal magnetization, is known as longitudinal relaxation and is given the symbol T_1.

The nuclei in an assembly of atoms are extremely well isolated from their surroundings, so much so that in certain especially pure solid samples values of T_1 of 1000 seconds have been recorded. There is in fact no reason why longitudinal relaxation should take place at all in the absence of stimulating radiation at the nuclear resonant frequency. In liquid samples however, the values of T_1 for spin $\frac{1}{2}$ nuclei vary typically from about 50 to 0·5 seconds and we have therefore to postulate that fluctuating magnetic fields exist within those samples which cause relaxation. Consideration of the nature of liquids shows that we should expect such fields to exist, since the molecules are undergoing Brownian motion, i.e. diffusing and rotating randomly so that the nuclear magnets in the system are continually moving relative to each other and so produce wildly fluctuating magnetic fields at all points within the sample. Over the long term these average to zero but at any instant will contain a component of random intensity and phase at the nuclear resonant frequency, and it is this which effects interchange of nuclear energy with the rest of the system and gives rise to the T_1 relaxation mechanism. Since the internal field also has random phase throughout the sample it will bring about a random distribution of spins on the conical surface (Fig. 36b), if B_1 is switched off, and so it also controls the T_2 relaxation mechanism. For this reason T_1 and T_2 are usually about equal in liquids. Note that the random phase T_2 relaxation field exists at all times even when B_1 is switched on. There is thus even at resonance, a tendency for spins to get out of phase with B_1 either going in advance or being retarded and this reduces the sharpness of the resonance. The line-width is in fact proportional to $1/T_2$.

B_1 will cause a net absorption of energy by the system since excess nuclei in the low energy state will be promoted to the high energy state. This will deplete the excess low energy population which gives rise to the signal and the signal would eventually be expected to vanish, except that the T_1 mechanism tends always to restore equilibrium. If B_1 is made very large however, then the internal relaxation field can no longer cope, the system is said to saturate and the signal intensity diminishes.

It is also worth considering for a moment the relaxation processes in solids. Here there is no Brownian motion so that there is no random relaxing field and T_1 is very long. The magnitude of T_1 is determined by the concentration of paramagnetic impurities in the solid lattice and for this reason is also known as spin-lattice relaxation. Because of the lack of Brownian motion, neighbouring spins come directly under the influence of each others magnetic fields. This consists of two components, (a) a static one due to the μ_z component of the nuclear magnet along the magnetic field axis and of a magnitude and direction determined by the spin state of the nucleus, and (b) a rotating one due to the transverse field component μ_{xy} of the precessing spin. The rotating component forms a means for the direct interchange of energy between spins and such interchange occurs frequently though with no net change in energy of the nuclear system and therefore no effect on T_1. The lifetime of the individual spin states is shortened drastically by this process and the uncertainly principle dictates that resonances will be broad. T_2 is indeed very short in solids, often in the region of microseconds and line-widths are measured in kilohertz. T_2 is sometimes referred to as spin-spin relaxation.

The static component of the fields of the nuclear magnets determines the line shape. If the nuclei occur in close pairs then for two like, spin $\frac{1}{2}$ nuclei, either will experience one of two magnetic fields depending upon whether the other has the orientation with $m = +\frac{1}{2}$ or $-\frac{1}{2}$. The resonance appears as a broad doublet. If the nuclei occur in close triangular clusters, then individuals experience one of three magnetic fields depending upon whether the spin orientations of the other two give $\Sigma m = +1$, $\Sigma m = 0$ or $\Sigma m = -1$. In this case the resonance appears as a broad $1:2:1$ triplet. Weaker interactions with more distant nuclei lead to further multiplicity which is not resolved but contributes further line broadening. The shape of these broadened multiplets can be used to determine internuclear distances which in the case of protons may be difficult to obtain using other techniques.

4.2   The Bloch equations

The above ideas have been incorporated into a series of equations which describe exactly the behaviour of an assembly of nuclear magnets. We will outline the derivation of the equations here and show how the result can be used to obtain expressions for line-shape, line-width, and saturation behaviour in terms of the two relaxation times. Thus for a single nuclear magnet we have, from (1.1):

$$\mu = \gamma(I\hbar) \tag{4.1}$$

If we transform this to vector notation then the vector equation describes the orientation of the magnetic moment μ in space:

$$\bar{\mu} = \gamma(\bar{I}\hbar) \tag{4.2}$$

The way that this orientation changes with time is then given by

$$\frac{d\bar{\mu}}{dt} = \gamma \frac{d}{dt}(\bar{I}\hbar) \tag{4.3}$$

The rate of change of momentum $(\bar{I})$ is a force and the force on a magnetic moment μ in a field B_0 is given by $\bar{\mu} \times \bar{B}_0$ so that we can write:

$$\frac{d\bar{\mu}}{dt} = \gamma(\bar{\mu} \times \bar{B}_0) \tag{4.4}$$

If $\bar{M}$ is the sum of all magnetic moments $\bar{\mu}$ we can replace $\bar{\mu}$ in the above equation by $\bar{M}$ which then relates to an assembly of nuclear magnets. Equation (4.4) can be rewritten as three equations in cartesian co-ordinates to describe the behaviour of M in each of the three x, y, z directions, with the applied field B_0 in the Z direction as before. We obtain:

$$\frac{dM_x}{dt} = \gamma B_0 M_y \tag{4.5a}$$

$$\frac{dM_y}{dt} = -\gamma B_0 M_x \tag{4.5b}$$

$$\frac{dM_z}{dt} = 0 \tag{4.5c}$$

These describe the Larmor precession of $\bar{M}$ with a vector M_{xy} rotating in the xy plane and with a steady component M_z in the B_0 field direction (Fig. 37).

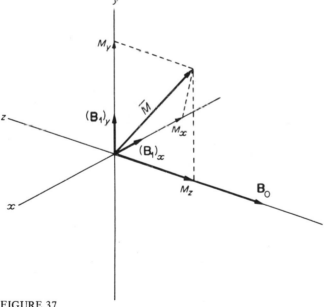

FIGURE 37
Diagram showing the meaning of the various vectors and fields used in formulating the Bloch equations. B_0 causes precession of $\bar{M}$ components M_x and M_y, $(B_1)x$ causes precession of M_z and M_y and $(B_1)y$ causes precession of M_x and M_z.

This however does not describe our system completely, because we know that M_{xy} will decay to zero as the nuclei spread out around the z axis, and that if we alter B_0 then M_z will take time to adjust to its new value. Thus we add a term to each of (4.5a) to (4.5c) to describe these changes. Equation (4.5c) becomes:

$$\frac{dM_z}{dt} = -\frac{M_z - M_0}{T_1} \qquad (4.6c)$$

where M_0 is the equilibrium value of M_z and $M_z - M_0$ is the displacement from this equilibrium value. The solution of this equation is:

$$(M_z - M_0)_t = (M_z - M_0)_0 \exp(-t/T_1)$$

and thus describes a relaxation process with characteristic time T_1. The other two equations are similarly modified, though since the equilibrium value of M_x and M_y are zero we get a slightly different form.

$$\frac{dM_x}{dt} = \gamma B_0 M_y - \frac{M_x}{T_2} \qquad (4.6a)$$

$$\frac{dM_y}{dt} = -\gamma B_0 M_x - \frac{M_y}{T_2} \qquad (4.6b)$$

which now allows M_{xy} to decay to zero after a sufficiently long interval.

The next step is to introduce a rotating magnetic field B_1 of angular frequency ω in the xy plane. This is resolved into two sinusoidally oscillating components along the x and y axis.

$$(B_1)_x = B_1 \cos \omega t \qquad (4.7a)$$

$$(B_1)_y = B_1 \sin \omega t \qquad (4.7b)$$

Each of these components will affect the nuclear magnetic vector in the same way as does the fixed field B_0. Thus for instance the effect of $(B_1)_y$ is to produce no change in M_y but to introduce a precession around the y axis, thereby introducing contributions to dM_x/dt and dM_z/dt. We can write a series of equations like (4.5a–c) to describe this motion and add them to equations (4.6). Thus the full equations become

Vector rotating around B_0	Relaxation terms	Precession around y axis	Precession around x axis	
$\dfrac{dM_x}{dt} = \gamma B_0 M_y$	$-\dfrac{M_x}{T_2}$	$-\gamma B_1 M_z \sin \omega t$	0	(4.8a)
$\dfrac{dM_y}{dt} = -\gamma B_0 M_x$	$-\dfrac{M_y}{T_2}$	0	$-\gamma B_1 M_z \cos \omega t$	(4.8b)
$\dfrac{dM_z}{dt} = 0$	$-\dfrac{M_z - M_0}{T_1}$	$-\gamma B_1 M_x \sin \omega t$	$-\gamma B_1 M_y \cos \omega t$	(4.8c)

The terms are arranged in columns depicting the contributions to the magnet motion of B_0, relaxation and B_1 respectively.

In order to solve equations (4.8) we effect a transformation to a new set of rotating axes. We define two quantities u and v by:

$$u = M_x \cos \omega t \quad M_y \sin \omega t$$

$$v = -M_x \sin \omega t - M_y \cos \omega t$$

These are the components of the transverse magnetisation along the B_1 direction and at right-angles to it in the xy plane. It is these components which induce an emf in the receiver coil and give rise to the n.m.r. signal. A series of straightforward though protracted algebraic manipulations then give us a new set of equations giving the differentials du/dt, dv/dt, and dM_z/dt as functions of u, v, γ, B_1, T_1, T_2, M_0, ω, and B_0. γ and B_0 invariably appear together and can be expressed as ω_0, the nuclear angular frequency. $\omega_0 - \omega$ is then a measure of the displacement of the B_1 probe field frequency from resonance.

We then make the important simplifying assumption that we sweep very slowly through the resonance condition so that the differentials change very slowly and can be set equal to zero. This is called the slow passage approximation. The resulting non-differential equations can be easily solved and give:

$$u = M_0 \frac{B_1 T_2^2(\omega_0 - \omega_1)}{1 + T_2^2(\omega_0 - \omega)^2 + \gamma^2 B_1^2 T_1 T_2} \tag{4.9}$$

$$v = M_0 \frac{B_1 T_2}{1 + T_2^2(\omega_0 - \omega)^2 + \gamma^2 B_1^2 T_1 T_2} \tag{4.10}$$

The output of an n.m.r. spectrometer thus contains two separate pieces of information. It is necessary to select one or the other and it is for this reason that a phase sensitive detector is used to rectify the spectrometer output, so that by setting the correct phase relationships between B_1 and M_{xy} a d.c. output proportional to either the pure u or pure v signal can be obtained which varies with ω according to (4.9) or (4.10). A typical spectrometer output is shown in Fig. 38. Normally the pure absorption mode signal is used since its position is defined accurately by its maximum. It may be objected by the watchful that none of the lines in the spectra in the previous examples bears any resemblance to Fig. 38a. This arises because spectra are normally run rather too fast for the slow passage approximation to be obeyed and so we observe transient phenomena following the narrow lines, and the lines are unsymmetrical.

We can nevertheless derive a number of important facts about the behaviour of an n.m.r. resonance from equation (4.10).

If the term $\gamma^2 B_1^2 T_1 T_2$ is very much greater than 1, then near resonance, where $\omega_0 - \omega$ is small (4.10) can be written:

$$v = M_0 \frac{B_1 T_2}{\gamma^2 B_2^2 T_1 T_2}$$

which simplifies to:

$$v = M_0 \frac{1}{B_1 T_1} \qquad (4.11)$$

This condition arises if the relaxation times are long or if B_1 is large and results in the signal intensity decreasing as B_1 is increased, i.e. in saturation. $\gamma^2 B_1^2 T_1 T_2$ is known as the saturation term. The existence of saturation places a definite limit to the sensitivity of n.m.r. spectrometers.

If the saturation term is very much smaller than 1 then it can be neglected and (4.10) becomes:

$$v = M_0 \frac{B_1 T_2}{1 + T_2^2(\omega_0 - \omega)^2} \qquad (4.12)$$

The signal is thus proportional to B_1 if B_1 is small or relaxation is fast. The signals obtained at several spectrometer B_1 settings are shown in Fig. 39. Note that since saturation is most pronounced near the centre of the resonance when $\omega_0 - \omega$ is small, the lines appear to broaden at high power.

70

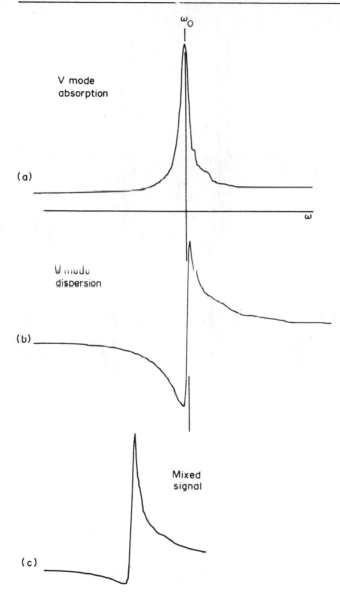

ω_O

V mode
absorption

(a)

ω

U mode
dispersion

(b)

Mixed
signal

(c)

FIGURE 38

(a) Absorption and (b) dispersion nmr resonances. (c) shows a mixed mode signal obtained by maladjusting the relationship between B_1 and nuclear phase at the detector.

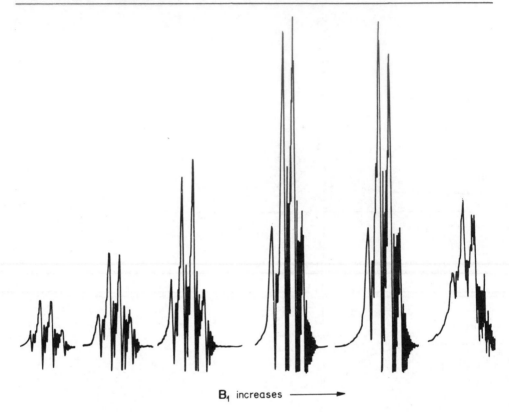

B₁ increases ⟶

FIGURE 39
The effect of progressively increasing B_1. The voltage input to the spectrometer was increased by two times for each step.

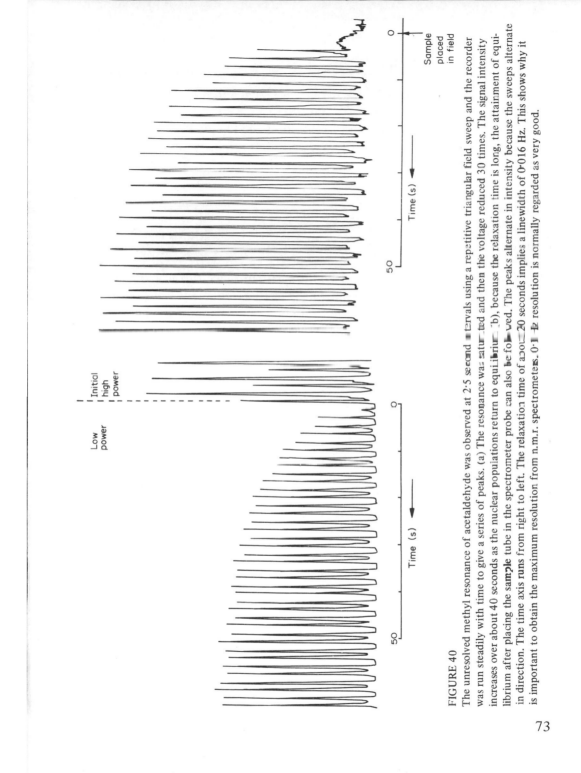

FIGURE 40

The unresolved methyl resonance of acetaldehyde was observed at 2·5 second intervals using a repetitive triangular field sweep and the recorder was run steadily with time to give a series of peaks. (a) The resonance was saturated and then the voltage reduced 30 times. The signal intensity increases over about 40 seconds as the nuclear populations return to equilibrium (1b), because the relaxation time is long, the attainment of equilibrium after placing the sample tube in the spectrometer probe can also be followed. The peaks alternate in intensity because the sweeps alternate in direction. The time axis runs from right to left. The relaxation time of about 20 seconds implies a linewidth of 0·016 Hz. This shows why it is important to obtain the maximum resolution from n.m.r. spectrometers. 0·1 Hz resolution is normally regarded as very good.

The line-shape predicted by (4.12) is Lorenzian and has detectable intensity at positions well away from resonance.

Equation (4.12) can also be used to predict the line-width of the resonance. If we define line-width as the width at half-intensity then the half-intensity points occur when:

$$1 + T_2^2(\omega_0 - \omega)^2 = 2$$

i.e. when

$$T_2^2(\omega_0 - \omega)^2 = 1$$

or

$$\omega_0 - \omega = \pm \frac{1}{T_2}$$

The separation of the half-intensity points is then twice this in radians. It is however more usual to express line-widths in Hz (frequency = $\omega/2\pi$) giving $\nu_{\frac{1}{2}} = 1/\pi T_2$ where $\nu_{\frac{1}{2}}$ represents the frequency separation of the half-height points.

Two simple ways in which T_1 can be estimated are indicated in Fig. 40.

4.2.1 *Wiggle beats*

The beat patterns which follow the fast traversal of a resonance and are often called wiggle beats are a marked feature of n.m.r. spectra. They can be predicted from a full treatment of the Bloch equations but can also be predicted qualitatively. Before we can do this however we have to understand how a phase sensitive detector works. This is set out diagrammatically in Fig. 41. A change-over relay is driven by a square wave so that the output terminals are connected to the input in one sense during a positive half-cycle and in the opposite sense during the negative half-cycle. If a signal of the same frequency as the drive waveform is applied to the input this is rectified by the regular switching of the relay and a dc output is obtained if the two inputs are in phase or 180° out of phase. In the latter case the sign of the output is reversed. If the phase relationship of the two inputs is other than 0° or 180°, an alternating component is obtained, but if this is integrated (or filtered) a dc output results but whose magnitude and sign depends upon the relative phases of the inputs. When the phase difference is 90° there is no dc output. If the input is the signal picked up from our nuclear magnets and the drive (or switching) wave form is obtained from the input to the coils producing B_1, we can then arrange their phases so as completely to reject either the u

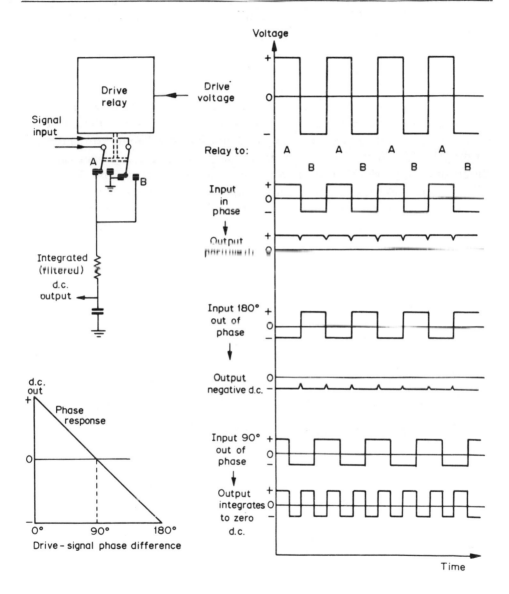

FIGURE 41

The phase sensitive detector. The output magnitude depends both upon the phase of the input relative to the drive and upon the magnitude of the input.

or the v mode signal. Electronic switching is of course required at the radio frequencies normally used for n.m.r. By suitable arrangement of phase the absorption mode signal can be made either positive or negative going, so that we cannot conclude from the trace whether in fact the n.m.r. effect is an absorption or an emission phenomenon.

Another interesting property of the phase sensitive detector is that if the input is not of the same frequency as the switching wave form, the output fluctuates at the difference frequency and does not contribute to the dc output. Thus the device effectively rejects all unwanted frequencies and detects only those fluctuations in the nuclear assembly caused by B_1.

We can now take up our main topic again and see how the beat patterns arise following sharp lines. We note that they only arise in the fast passage case where the line is traversed in a time short compared with the relaxation time. As we sweep the frequency of our B_1 probing field towards resonance we start to tip the cone of nuclear vectors away from the z axis and so produce a

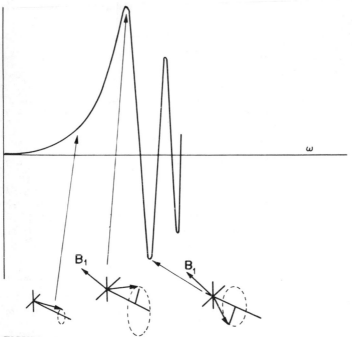

FIGURE 42

Showing the build-up of M_{xy} which occurs as the resonance condition is reached, followed by the very rapid change of phase (and sign of output) as the nuclear and B_1 frequencies diverge again.

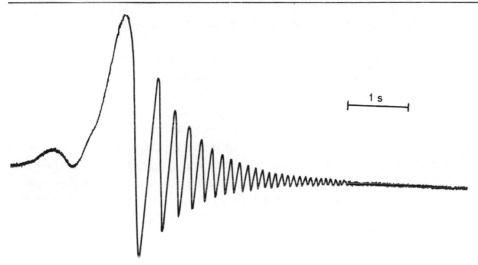

FIGURE 43

A typical wiggle pattern (^{7}Li in aqueous LiCl)

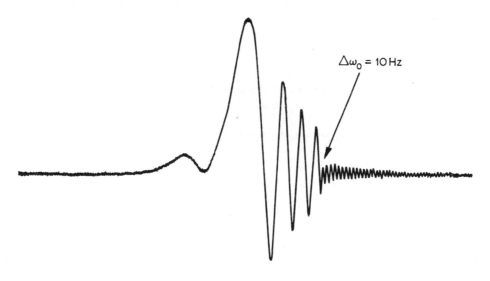

FIGURE 44

The spectrometer frequency was changed suddenly by 10 Hz during a wiggle beat response. The wiggle frequency changes and the intensity drops due to the poor high frequency response of the spectrometer.

signal. This process continues with increasing effectiveness till we reach resonance at ω_0. At this point the magnetisation in the xy plane, M_{xy} is in step with the B_1 vector in the phase relation to give a maximum output. If the B_1 frequency continues to change at the quite usual rate of 5 Hz/s, then within 0·33 s the B_1 vector will have moved 180° away from the nuclear vector and the spectrometer output will have reversed sign. This time is insufficient for M_{xy} to have relaxed appreciably so that the intensity will still be high (Fig. 42). Thereafter the output will invert more and more rapidly as the nuclear and B_1 frequencies diverge, while at the same time the intensity will fall as M_{xy} relaxes to zero. The persistence of the wiggle beat envelope gives some idea of the length of T_2 though it is also determined by the magnetic field inhomogeneity and the band-width of the spectrometer circuits (or frequency response of its dc amplifiers) and can never be used actually to measure T_2. Fig. 43 shows an actual wiggle pattern and Fig. 44 the effect of a sudden jump in spectrometer frequency made during a wiggle response. This demonstrates both the change in wiggle frequency which occurs and the sudden drop in output due to the falling spectrometer frequency response.

Wiggle beats are by no means mere ornament to our spectra but have two important uses. If we have two lines close together each produces its separate wiggle pattern and these overlap and interfere. The frequency of the wiggle beat following one resonance at any instant is given by:

$$\nu_{\text{wiggle}} = \nu_{\text{nucleus}} - \nu_{B_1}.$$

while for the second resonance with different nuclear frequency we have:

$$\nu_{\text{wiggle}}^1 = \nu_{\text{nucleus}}^1 - \nu_{B_1}$$

subtracting:

$$(\nu - \nu^1)_{\text{wiggle}} = \nu_{\text{nucleus}} - \nu_{\text{nucleus}}^1$$

Now the term on the right is the frequency chemical shift between the two nuclei which is of course constant, so that we can expect to see a constant frequency beat component in the wiggle pattern whose frequency is equal to the Hz chemical shift between the lines. The frequency separation of unresolved lines can be measured accurately in this way or the distortion of a wiggle pattern may often suggest the presence of unresolved splittings (Fig. 45).

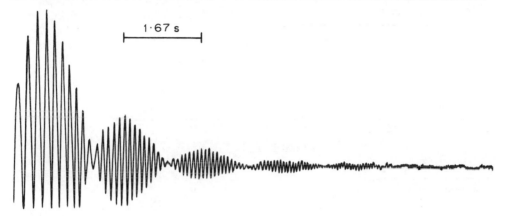

FIGURE 45

The beat pattern following a fast traversal of the ^{19}F resonance of $(CF_3S)_2NH$. There is weak coupling between fluorine and the proton so that the resonance is a doublet. This cannot be resolved but is obvious from the wiggle pattern. The beat interval gives ^4J (F-H) as 0.6 Hz.

The second use follows from this observation. If the magnetic field varies over the sample, i.e. is inhomogeneous, then the lines in our spectrum will be poorly resolved. Individual lines will be broadened and can be thought of as being made up of several overlapping lines originating from different parts of the sample. These will give interfering wiggle beat patterns and so the exponential wiggle decay will be distorted. This gives a means of rapidly testing the field inhomogeneity, and adjustments to the shim currents can be made until the correct shape of pattern is obtained. The acetaldehyde multiplets are commonly used for this purpose and the beat pattern generated by a single passage of the resonance can be observed to persist for up to 20 seconds on a well adjusted spectrometer.

4.2.2   Relaxation and sample viscosity

The internal fluctuating magnetic field which causes relaxation owes its existence to Brownian motion and the time-scale of the motion is one factor which determines its effectiveness. Since Brownian motion is random in character its time-scale is characterised by a somewhat loosely defined term, the correlation time τ_c. This is defined as the time taken on average for a molecule to diffuse a distance equal to its own dimension or to rotate through one radian. τ_c is typically 10^{-11} sec in liquids of low viscosity. This corre-

79

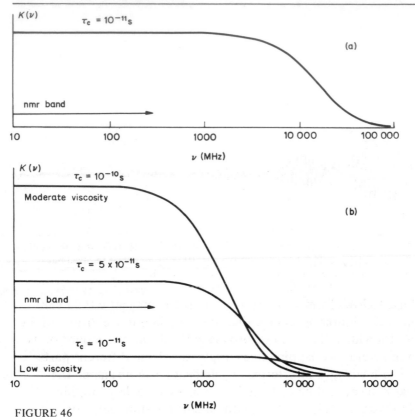

FIGURE 46

Intensity of fluctuations in magnetic fields in a liquid sample due to Brownian motion, as a function of frequency: (a) typical liquid; (b) liquids of different viscosities.

sponds to a frequency of 10^5 MHz. It has been shown that the frequency spectrum of randomly fluctuating fields such as exists within n.m.r. samples is essentially that of 'white' noise and contains all frequencies less than $1/\tau_c$. The field intensity at any frequency, $K(\nu)$, is given by:

$$K(\nu) \propto \frac{2\tau_c}{1 + 4\pi^2\nu^2\tau_c^2}$$

This fuction is plotted in Fig. 46a and it will be seen that over the n.m.r. frequency band the relaxation field intensity is constant. The Debye theory of electric dispersion shows that for a spherical molecule rotating in a liquid the correlation time and viscosity are related by:

80

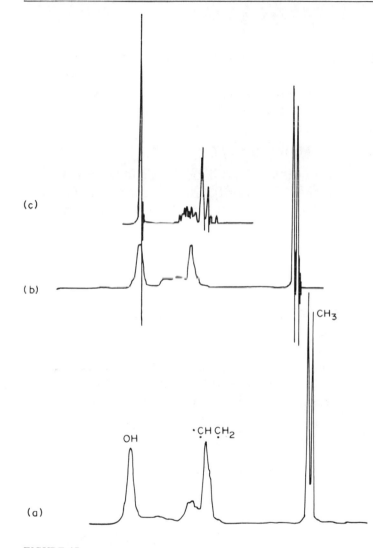

FIGURE 47

Proton spectrum of propylene glycol, $CH_3CH\,OH\,CH_2\,OH$: (a) neat viscous liquid; (b) 16 per cent solution in deuterochloroform. The methyl doublet is much better resolved and the well-developed wiggle pattern shows that the relaxation time is longer. The remaining resonances are still broad.
(c) A drop of HCl is added which promotes hydroxy proton exchange, destroys all spin coupling to these protons and they and the CH CH₂ backbone protons all become better resolved. Some field drift has occurred between spectra.

81

$$\tau_c = \frac{4\pi a^3}{3k} \frac{\eta}{T}$$

where η is the viscosity, T is the temperature and a is the radius of the sphere. Thus if we vary the viscosity of a sample or vary its temperature we will alter τ_c. The effect of increasing τ_c is shown in Fig. 46b and for the viscosities of liquids normally used as solvents its effect in the n.m.r. band is simply to increase $K(\nu)$ proportionately, i.e.:

$$K(\nu) \propto \tau_c \propto \eta/T$$

Thus in order to obtain long relaxation times, T_1 and T_2, and the narrowest resonances, it is necessary to keep the viscosity low. The spectrum of a viscous liquid is compared in Fig. 47 with its relatively non-viscous solution to emphasise this point.

Viscosity is of course not the only factor to affect the intensity of the relaxation field and this depends also upon the distance between the nucleus and those nuclear magnets producing the relaxation field and upon their magnetic moments. The proton has the largest magnetic moment and so the relaxation field tends to be a maximum in hydrogen-containing substances. Other nuclei should therefore have rather longer relaxation times but this is found to be true only for spin $\frac{1}{2}$ nuclei. For nuclei with $I > \frac{1}{2}$ the relaxation times are always shorter than would be expected on the basis of magnetic relaxation alone, often by factors as large as 10^8. We therefore have to consider a second relaxation mechanism for these nuclei.

4.3   Electric quadrupole relaxation

Nuclei with $I > \frac{1}{2}$ possess a quadrupole moment Q which arises because the distribution of charge in the nucleus is not spherical but ellipsoidal, i.e. the charge distribution within the nucleus is either slightly flattened (oblate — like the earth at its poles) or slightly elongated (prolate — like a rugby ball). Cross-sections of the charge of two such nuclei are shown in Fig. 48 with the departure from spheroidal form much exaggerated. The disposition of the electric quadrupole is also suggested. If the electric fields due to external charges vary across the nucleus then the torque on each dipole component of the quadrupole is different and a net torque is exerted on the nucleus by the electric field as well as by the magnetic fields present. Electric field

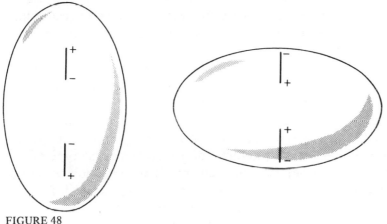

FIGURE 48
Cross-sections of the charge distributions in quadrupolar nuclei.

gradients exist at atomic nuclei due to asymmetries in the spatial arrangement
of the bonding electrons. The Brownian tumbling of the molecule causes the
direction of the resulting electric quadrupole torque to vary randomly
around the nucleus in exactly the same way as does the torque due to the
magnetic relaxation field. Rotating electric torque components thus exist at
the nuclear resonant frequency which can also cause interchange of energy
between nucleus and the rest of the system (T_1 mechanism) and random-
isation of nuclear phase (T_2 mechanism).

The quadrupolar mechanism is a highly effective means of relaxation of
many nuclei, and often can be regarded as the sole source of relaxation. Its
magnitude depends upon a large number of factors so that the observed
relaxation times vary over a very wide range. A simplified form of the quadrupole relaxation equation is given below which emphasises the importance
of nuclear properties I and Q and of the maximum field gradient $- d_2V/dz^2$.
T_1 and T_2 are equal:

$$\frac{1}{T_1} = \frac{1}{T_2} \propto \frac{(2I + 3)Q^2}{I^2(2I - 1)} \left(\frac{d_2V}{dz^2}\right)^2 \tau_c$$

Nuclear quadrupole relaxation thus increases rapidly in effectiveness as
we move from observing nuclei with small Q to those with large Q, though
this is to some extent offset by the factor involving I. This occurs because

there is a tendency for large Q to be associated with large I and because the value of $(2I + 3)/[I^2(2I - 1)]$ falls off rapidly with increasing I. For a given nuclear species the relaxation time is primarily determined by the electric field gradient. This depends mainly upon the electronic symmetry around the nucleus (not upon the symmetry of the molecule as a whole though the two are often linked). A uniform spherical orbital produces zero electric field gradient as does any arrangement of localised charges with cubic symmetry. Thus nuclei of free ions and nuclei of atoms at the centres of strictly regular tetrahedral or octahedral structures formed from a single type of ligand, all possess near zero electric field gradients, long relaxation times and therefore narrow lines. At the other extreme, nuclei in atoms which form a single bond to other groupings have large electric field gradients, short relaxation times and broad, sometimes undetectable lines. The disposition of non-bonding pairs of electrons and of electrons forming π-bonds can however considerably reduce the electric field gradients at such terminal atoms. Table 4.1 on p 86 contains a number of examples in illustration. For comparison typical natural proton line-widths in a perfectly homogeneous field are 0·02 Hz.

Study of the table will indicate what line-widths are to be expected in various situations. Note that in nitrogen compounds the bonding to nitrogen is all important in determining the field gradient so that substitution of H by phenyl in $C_6H_5CH_2NMe_3^+$ produces very little change in line-width though the symmetry of the molecule as a whole changes markedly. The low field gradient at the isocyanide nitrogen is also remarkable.

'Terminal' seems to mean very different things in terms of electric field gradient to chlorine in PCl_3 and nitrogen in MeCN and indicates very different electronic structures around these two atoms. The high field gradient in PCl_3 may suggest that the amount of sp^3 hybridisation of the chlorine atoms is quite small.

At the same time it must be admitted that there is very little systematic correlation of line-widths with structure and while it is possible to differentiate between atoms in positions of low or of high symmetry little can be said at present about intermediate cases.

4.3.1   *Spin-spin coupling to quadrupole relaxed nuclei*

So far we have considered the effect that relaxation has upon line-width and saturation. Since T_1 relaxation involves changes in spin orientations we might also expect relaxation processes to modify the spin coupling patterns we observe. The relaxation times of spin $\frac{1}{2}$ nuclei are sufficiently long for us not normally to have to take this into account when analysing spectra, but the much shorter relaxation times of the quadrupolar nuclei do lead to considerable modification of the observed patterns.

If T_1 is long then the normal spin coupling effects are observed. Thus spin coupling of the methyl group to ^{11}B is observed in the adduct $MeNH_2BF_3$ though reference to Fig. 22 will show that the lines are rather broadened. At the other extreme, if T_1 is very short, the nuclear spins interchange energy and change orientation so rapidly, that a coupled nucleus interacts with all possible spin states in a short time. It can distinguish only an average value of the interaction and a singlet resonance results. This explains for instance why the chlorinated hydrocarbons show no evidence of proton spin coupling to the chlorine nuclei ^{35}Cl and ^{37}Cl, both with $I = 3/2$. At intermediate relaxation rates the coupling interaction is indeterminate and a broad line is observed. The line-shapes calculated for the resonance of a spin $\frac{1}{2}$ nucleus coupled to a quadrupole relaxed spin 1 nucleus such as ^{14}N are shown in Fig. 49. The shape of the spectrum observed depends upon the product $T_1 J$, where T_1 is the relaxation time of the quadrupole nucleus, since if the frequency defined by $1/(2\pi T_1)$ is comparable with the coupling constant in Hz then the coupled nucleus cannot distinguish the separate spin states. The situation is equivalent to attempting to measure the frequency of a periodic wave by observing only a fraction of a cycle. The resonance of the quadrupolar nucleus will of course be split into a multiplet by the spin $\frac{1}{2}$ nuclei, but each component line will have width $1/(\pi T_2)$ Hz.

A common example of lines broadened by coupling to quadrupolar nuclei is found in amino compounds. The protons on the nitrogen are usually observed as a broad singlet, a good example appearing in Fig. 27b. It is important to remember in this case that the relaxation time of the amino proton is unaffected and can cause normal splitting in vicinal protons bonded to carbon. In contrast the protons in the highly symmetrical ammonium ion give narrow resonances because the nitrogen quadrupole relaxation is slow (Fig. 20b).

TABLE 4.1
Line-widths of the resonances of some quadrupolar nuclei

Molecule	Symmetry around nucleus	Nuclear resonance line width (Hz)
Line-widths of the ^{14}N resonance in:		
The tetramethylammonium cation Me_4N^+	Strictly regular tetrahedron	7
The benzyltrimethylammonium cation $Me_3NCH_2C_6H_5^+$	Tetrahedral around nitrogen. The efg[††] is low and the environment must appear to the nitrogen as a nearly regular tetrahedron	12
The ammonium cation H_4N^+	Strictly regular tetrahedron	2[†]
The phenylammonium cation $H_3NC_6H_5^+$	Tetrahedral but one ligand very different from the other three	100
Liquid ammonia NH_3	Pyramidal but lone pair electrons give some tetrahedral character and the efg is similar to that of the previous ion	135[‡]
Methyl isocyanide MeNC	Linear but with very low efg at N. The arrangement of the electrons seems fortuitously to give an even lower efg than in the regular tetrahedron of NH_4^+	1[†]
Methyl cyanide MeCN	Terminal	170
The azide anion (end nitrogens) $:-\quad+\quad-:$ $:N=N=N:$	Terminal, field gradient reduced by lone pairs and π-bonding	35
Line-widths of the ^{35}Cl resonance in:		
Phosphorus trichloride PCl_3	Terminal	83000
The perchlorate anion ClO_4^-	Strictly regular tetrahedron	3

The relative values of the expression $(2I + 3)Q^2/[I^2(2I - 1)]$ for the above nuclei are $^{14}N = 0.025$, $^{35}Cl = 0.0083$, $^{27}Al = 0.007$, $^{11}B = 0.0017$ and $^{59}Co = 0.034$ ($e^2 \times 10^{-48}$ cm^4).

Molecule	Symmetry around nucleus	Nuclear resonance line width (Hz)
Line-width of the ^{27}Al resonance in:		
The hexaquoaluminium cation Al(H$_2$O)$_6^{3+}$	Regular octahedron	$\geqslant 3$
Triisobutyl aluminium AlBu$_3^i$	Trigonal	6000
Line-width of the ^{11}B resonance in:		
The tetrafluoroborate anion BF$_4^-$	Strictly regular tetrahedron	0.3
Trimethyl borate B(OMe)$_3$	Trigonal	25
Diborane B$_2$H$_6$	Distorted tetrahedron	~ 5
Line-width of the ^{59}Co resonance in:		
[Co(PF$_3$)$_4$]$^-$	Strictly regular tetrahedron	< 57
Co$_4$(CO)$_{12}$	The four cobalt atoms are arranged in a tetrahedron. One carries three carbonyl ligands and is therefore in a distorted octahedral environment. The other three carry two ligands and are bonded together by metal-metal bonds and carbonyl bridges. Their environment is thus seven coordinate. Two resonances are observed	7500

† These ^{14}N resonances were obtained on more modern spectrometers than the remainder and are probably less broadened by magnetic field inhomogeneities. The line-width of Me$_4$N$^+$ is certainly less than 7 Hz. All the resonances listed above will contain unknown broadening due to field inhomogeneities and so are slightly overestimated.

†† efg = electric field gradient.

‡ Corrected to allow for the very low viscosity of liquid ammonia. The measured value is 35 Hz.

Since quadrupole relaxation is sensitive to temperature and viscosity the line-shapes observed for coupled nuclei are altered by viscosity and temperature changes, an increase in temperature leading to *slower* relaxation and a better resolved multiplet. This fact is stressed since on a first encounter it appears to be contrary to ones expectation.

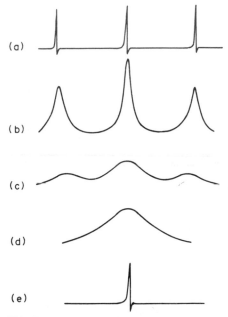

FIGURE 49

The resonance line-shape of a spin $\frac{1}{2}$ nucleus coupled to a spin 1 nucleus having various rates of quadrupole relaxation: (a) T_1 long; (b) $T_1 \approx 8/2\pi J$; (c) $T_1 \approx 3/2\pi J$; (d) $T_1 \approx 1/2\pi J$; (e) T_1 very short. The intensities are not to scale. The broadened lines are weak since the total area is constant.

4.4   N.M.R. spectra of exchanging systems

One of the most important contributions n.m.r. has made to chemistry is the insight it has given into the dynamic nature of many molecular systems. Spectroscopy based on higher frequency radiation has given mostly a static picture because the time-scale of many processes is slow relative to the frequency used. However, the lower frequencies used for n.m.r. and the smaller line-separations involved, coupled with the small natural line-widths obtained, means that many time-dependent processes affect the spectra profoundly. We have of course just discussed one such process, namely quadrupole relaxation and how it affects spin coupling patterns.

If two atoms of the same element in a system have different chemical environments then their nuclei (assuming these are the same species) will give rise to two chemically shifted resonance lines. If the atoms interchange position occasionally, their lifetime τ in any one environment is long and the spectrum is harldy altered. On the other hand if the lifetime is exceptionally short then the spectrometer can only distinguish an average environment and a singlet is observed. At intermediate lifetimes a broadened line is obtained whose shape is determined by $\tau \nu_0 \delta$, where δ is the ppm chemical shift between the resonances in the absence of exchange.

An example of this behaviour is found with the N-methyl resonance of N,N-dimethylformamide, $Me_2N.CHO$. This is a doublet at 25° because rotation around the C-N bond is restricted since the nitrogen lone pair electrons form a partial double bond to the carbonyl carbon, and the two methyl groups lie in different chemical environments (Fig. 50). If the compound is heated the rate of rotation increases and the line-shapes change in the way shown in the figure. Analysis of the curves gives the value of τ at each temperature and thence the activation energy for hindered rotation of the N-methyl groups. It is possible to observe these changes, albeit somewhat crudely, if a heated sample is placed in the coil of certain n.m.r. spectrometers and a series of spectra is swept rapidly as the sample cools.

Line-widths and spectral shapes change quite profoundly with temperature, and for this reason considerable effort has been put into the determination of exchange rates from spectral shapes, so that activation parameters can be obtained from Arrhenius type temperature-rate plots. In order to do this of course a quantitative theory of the method is required. We do not intend to discuss this here but merely give the results for slow and fast rates of exchange. If exchange rates are required in the intermediate case it is normal to compute the line-shape using a digital computer and find a best fit to the experimental line-shape.

If we consider exchange between two sites A and B where the residence times on each site are τ_A and τ_B and the relaxation times are T_{2A} and T_{2B} respectively, then the apparent relaxation time T_2' is reduced as follows:

For slow exchange:

$$\frac{1}{T_{2A}'} = \frac{1}{T_{2A}} + \frac{1}{\tau_A}$$

$$\frac{1}{T_{2B}'} = \frac{1}{T_{2B}} + \frac{1}{\tau_B}$$

i.e. two broadened lines are obtained. The broadening arises because the leaving spin does not necessarily have the same orientation as the arriving spin so that extra relaxation is introduced.

For fast exchange we define the proportions of A and B sites as P_A and P_B where:

$$P_A = \frac{\tau_A}{\tau_A + \tau_B} \qquad\qquad P_B = \frac{\tau_B}{\tau_A + \tau_B}$$

then

$$\frac{1}{T_2'} = \frac{P_A}{T_{2A}} + \frac{P_B}{T_{2B}} + P_A^2 P_B^2 (2\pi\nu_0\delta)^2(\tau_A + \tau_B)$$

giving a single broadened line of resonant frequency:

$$\nu' = P_A\nu_A + P_B\nu_B$$

at the weight averaged shift position. These two expressions can be used to gain a rough idea of exchange rates in the two limiting cases though even in this region, experience has shown that computer fitted line-shapes give the most reliable data. This is perhaps not surprising since when one measures a line-width at half-height one is using only three data points, whereas a line can be regarded as being made up of many data points which the simple process ignores.

Variable temperature n.m.r. spectroscopy has been used extensively to study many dynamic processes. The rates of exchange involved and line-separations vary considerably, so that some samples require cooling rather than heating to reach the temperature where the exchange rate affects the form of the n.m.r. spectra. Because of the importance of this type of work many spectrometers are equipped with devices for cooling or heating the sample.

In such spectrometers the sample is placed in a Dewar-jacketed probe and heated or cooled nitrogen gas passed through the probe. Typically, temperatures in the range $+ 200°C$ to $- 130°C$ can be achieved though temperature control is never so good as in a conventional thermostatic bath, due mainly

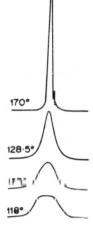

FIGURE 50

The N-methyl resonance of N, N-dimethyl-formamide neat liquid at different temperatures. Rotation of the Me₂N groups around the CN bond is slow at room temperature due to partial double-bond character in the bond.

At high temperatures the barrier to rotation is overcome and the two methyl groups see the same average environment. (After Bovey.)

91

to the low thermal capacity of the heat exchange medium. For the same reason one cannot guarantee that the whole of the sample is at the same temperature. Despite these drawbacks, much interesting information about exchange processes has emerged using variable temperature n.m.r. spectroscopy. Studies have for instance been made of the hindered rotation which occurs around the carbon-carbon bond in highly halogenated ethanes, of the flexing of saturated rings and of the exchange of ligands on transition metal complexes. Several further examples will be described in detail in Chapter 7.

N.M.R. can also be used to study protolysis reactions. In a pure alcohol, RCH_2OH, the methylene protons are coupled both to protons in R and to the OH group. Thus in the spectrum of pure ethanol (Fig. 51a) the hydroxyl and $.CH_2$ resonances are broadened due to this coupling. The coupling is not fully resolved except in very pure samples because of slow interchange of hydroxide protons. Addition of a drop of hydrochloric acid promotes faster exchange and the loss of all coupling interaction since the hydroxide protons are replaced frequently with random spin orientation. The resulting dramatic line-narrowing is seen in Fig. 51b. The exchange rates of the protons in such systems have been studied as a function of acid concentration by analysing the line-shapes obtained.

The spectra of ethanol also illustrate the effects upon n.m.r. spectra of chemical equilibria involving fast exchange. The hydroxy group of the alcohol is extensively inter-molecularly hydrogen-bonded. The degree of hydrogen bonding is not however 100 per cent so that the hydroxyl protons can be thought of simply as existing in two environments, one hydrogen-bonded and one non-hydrogen-bonded. Two resonances might be observed but for the fact that there is rapid interchange of hydroxyl roles, and a single line is observed at a position determined by the concentration-weighted average of the shifts in the two environments. If the degree of hydrogen bonding is altered then the resonance should move. A change in hydrogen bonding is brought about by diluting the alcohol in an organic solvent and a large hydroxyl shift is observed in Fig. 51c. The more extensively hydrogen-bonded neat alcohol has the lower field hydroxyl resonance. (The methyl and methylene resonances are affected relatively little.) Hydrogen bonding results in a low field shift possibly because of an electric field effect.

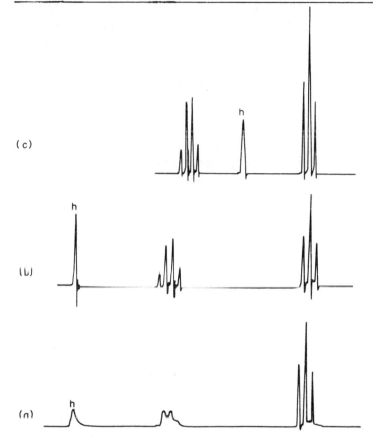

(c)

(b)

(a)

FIGURE 51

(a) 60 MHz proton spectrum of neat ethanol. The methyl protons give rise to the triplet at high field and the methylene protons give the quartet near the centre of the spectrum. Hydroxyl exchange is slow and the methylene quartet is broadened by partially collapsed spin coupling to the hydroxyl proton. This likewise gives a broadened singlet at low field labelled h. (b) A drop of hydrochloric acid markedly increases the rate of hydroxyl proton exchange and all vestiges of coupling between hydroxyl and methylene protons are destroyed. The hydroxyl (h) and methylene resonances sharpen dramatically. (c) A dilute (~ 7%) solution of ethanol in deutero chloroform. The solvent breaks up the intermolecular hydrogen bonds that exist in the neat liquid and the hydroxyl resonance (h) moves 2·8 p.p.m. upfield.

93

Similar changes to the above are observed in the spectra of propylene glycol (Fig. 47).

The hydrated proton resonates at very low field, and the addition of acid to water results in fast proton exchange between solvent and hydrated proton and a low field shift of the water resonance. This phenomenon has been used to study the dissociation of strong acids.

The destruction of spin coupling by exchange requires a further word. In the example given, the exchange involved bond breaking and replacement of one proton by another of not necessarily the same spin orientation. Thus the exchange can be regarded as leading to apparent nuclear relaxation. Exchange not involving bond rupture can however also affect spin coupling. If the coupling constant between two nuclei in a molecule is different for two conformations of the molecule, then if conformational interchange occurs the coupling will not be destroyed. If the exchange is slow, two molecular species will be present and two multiplet sets will be seen, one corresponding to each conformation. If exchange is fast then the corresponding lines of each multiplet set will be averaged and a single multiplet set will be obtained exhibiting average coupling constants and an average shift.

An important example of this is the spectrum of the ethyl grouping. The six vicinal coupling constants existing at any instant between the three methyl protons and the two methylene protons will be different since all the dihedral angles will be different; the chemical shifts of all the protons may also be different and a complex spectrum should result. Rapid rotation of the methyl groups however averages all the different couplings over a Karplus curve and also averages the shifts, so that only one coupling constant is observed. The three methyl protons are all magnetically equivalent and so are the two methylene protons. In considering equivalence of nuclei such rapid molecular motion should always be taken into account.

An important exception where rotation does not usually lead to complete averaging is found for methylene protons which are bound to a carbon atom which is bonded to another carbon atom carrying three different substituents, i.e. $RCH_2-CR_1R_2R_3$. The presence of the substituent R on the methylene carbon means that the methylene protons are never in exactly the same environment whatever the conformation and an AB spectrum results. Note that the methylene protons may remain inequivalent if R_1 is a second CH_2R group. Thus the presence of an asymmetric carbon is not essential for this type of inequivalence to occur.

Slow chemical reactions can of course also be followed using n.m.r., though in this case a series of normal spectra are obtained over a period of time with no evident exchange perturbation, except that the relative line intensities vary and some lines may disappear and new ones appear.

4.5   Double resonance experiments

We have so far discussed relaxation caused by the internal fluctuating magnetic and electric fields and by exchange processes. We can also effectively cause rapid relaxation if we irradiate a group of nuclei at their resonant frequency and so stimulate upward and downward nuclear transitions. We can do this by connecting an extra rf transmitter to the spectrometer and can then observe the effect that irradiating one resonance has upon the signals obtained from other nuclei in our sample. We find that certain signals are altered or perturbed by this procedure and that the affected nuclei are coupled in some way to the irradiated nuclei. This procedure is known as a double resonance experiment and has led to the performance of some most elegant n.m.r. investigations of chemical systems. The experiment can be homonuclear, i.e. among like nuclei, or heteronuclear, i.e. one nuclear species is observed while another is double irradiated. Nuclei can be coupled in several ways: via the bonds, i.e. spin-spin coupling; via chemical exchange; or via the through-space relaxation field. Each coupling path leads to different double resonance effects and we shall discuss them separately.

4.5.1   *Double resonance and spin-spin coupling*

The double irradiating field is applied in the same way as the B_1 field, i.e. in coils which give a rotating magnetic vector, which we will designate B_2. If B_2 is the magnitude of the irradiating field and γ is the magnetogyric ratio of the nucleus irradiated, then we can define a frequency ν_2 such that $\nu_2 = \gamma B_2/2\pi$.

Irradiating field strength is usually expressed in terms of ν_2, and may be of the order of one or less Hz, when single lines in spin multiplets may be selectively irradiated without affecting other lines (low power or tickling irradiation), or may be several Hz when all the the lines in a multiplet may be affected simultaneously (high power irradiation). In this case $\nu_2 \approx J$.

If we use low power double irradiation then we find that in order to observe any effect we have to tune our B_2 transmitter close to the frequency of an individual line, the lower the power the more exact must be the tuning. The rotating component of B_2 is then stationary relative to the irradiated nuclei. We can think of it as causing a precession of the irradiated nuclei at frequency ν_2 which modulates the coupling to other nuclei in the molecule and so splits a proportion of the lines of their multiplets into doublets of spacing ν_2. The lines which are split are those in the energy level diagram (Fig. 28) which have energy levels in common with the irradiated transition. For example, irradiation of C_1 perturbs $A_1B_1A_3B_3$. Thus a few low power experiments can be used to relate the lines in an observed spectrum to an energy level diagram and so find the relative signs of coupling constants among the nuclei.

A second and spectacular use of the low power technique is to obtain the spectra of nuclei which might not be observable with a given spectrometer. For instance most spectrometers will be able to produce a proton spectrum of the ammonium ion NH_4^+, which is a $1:1:1$ triplet, but very few can produce the ^{14}N spectrum, which is a quintet. The proton spectrometer can be used to produce the ^{14}N quintet with the addition only of an auxiliary transmitter operating at the ^{14}N frequency. No ^{14}N receiver or sample coil replacement is required.

The experiment is carried out by adjusting the spectrometer so that it is continually tuned to one of the proton resonances. The field must of course be very stable to allow us to do this. We obtain a constant spectrometer output from the protons. We then apply weak irradiation near the ^{14}N frequency using the auxiliary transmitter and sweep its frequency. Every time

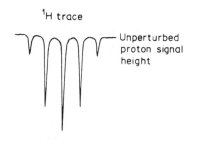

¹H trace

Unperturbed
proton signal
height

¹⁴N frequency

FIGURE 52
INDOR spectrum of the ^{14}N resonance of NH_4^+ in acidic solution obtained by observing one line of the proton resonance while sweeping a double irradiation transmitter through the ^{14}N resonances. $^1J(^{14}N-^1H) = 52.6$ Hz. (After Gilles and Randall.)

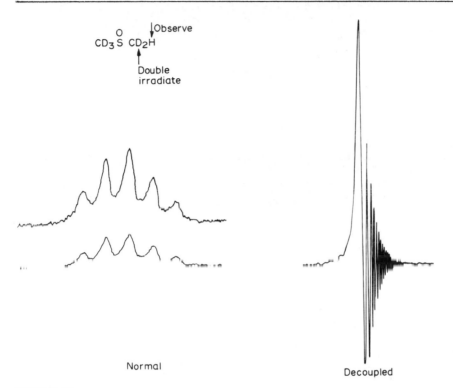

O
CD₃ S CD₂H
↓Observe
↑
Double
irradiate

Normal

Decoupled

FIGURE 53

The ^{1}H spectrum of pentadeutero dimethyl sulphoxide. The proton resonance of the CD_2H group is split into a quintet by the two deuterons. The splitting is removed when the deuterium resonance is strongly irradiated.

the frequency corresponds to one of the ^{14}N quintet lines, part of the proton line is split into a doublet and the intensity of the central line is reduced. Thus if we sweep the ^{14}N frequency and the recorder simultaneously we will observe a change in the spectrometer output every time a ^{14}N line is traversed and a quintet will be drawn out. The proportion split out depends upon the proportion of ^{1}H spin states contributing to the ^{14}N line and so the quintet is drawn with the correct intensities (Fig. 52). This is called an INDOR experiment.*

* i.e. INternuclear DOuble Resonance

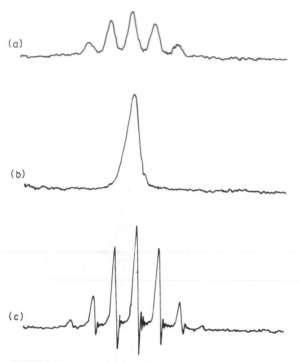

FIGURE 54

(a) The ^{31}P spectrum ($I = \frac{1}{2}$) of triethylphosphite $(CH_3CH_2O)_3P$. Both the methyl and the methylene protons are coupled to the phosphorus, the six methylene protons producing a septet and the methyls a decet. The P-to-methyl proton coupling is small and leads to unresolved broadening of the septet lines. The outer lines are lost in noise. (b) This is confirmed by irradiating the CH_2 methylene protons when only an unresolved decet remains due to the CH_3_C-P interaction. (c) If instead the methyl resonance is irradiated, the septet due to coupling to the methylene protons becomes well resolved since the methyl protons responsible for the broadening are decoupled.

High power double irradiation is applied near the centre of a multiplet whether there is a line there or not and its effect is to split all the lines arising from a coupled nucleus. The splitting is asymmetric and the lines nearest the centre of the multiplet are most intense. As ν_2 and the line splitting are increased the lines moving nearer to the centre of the multiplet crowd together while those moving away from the centre lose intensity. When ν_2 is large enough all the intensity is at the centre of the multiplet and a singlet is observed. The irradiated nucleus is said to have been decoupled. The condition for this is that $\nu_2 \gg J$. The technique is used for the simplifi-

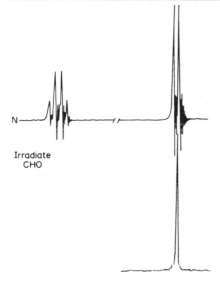

FIGURE 55

An example of high power homonuclear double irradiation. N shows the normal spectrum of acetaldehyde CH$_3$CHO and below it the spectrum of the methyl group run while the formyl proton was double irradiated. (After Bovey.)

cation of spectra and is most easily applied to the heteronuclear case where the nuclear frequencies are well separated, so that the powerful double irradiation transmitter cannot interfere with the spectrometer receiver system. Like nuclei can be decoupled if the resonances are separated by a frequency very much larger than the coupling constant between them (Figs. 53, 54, and 55).

Note that our initial assertion that high power double irradiation can be thought of as inducing rapid nuclear transitions so that spin coupling is lost is a considerable over-simplification of the double resonance effect.

4.5.2 Double resonance and exchange coupling

We have already seen that slow exchange of atoms between two sites A and B leads to line broadening. The areas of the lines of the exchanging nuclei however remain constant since the populations of the high and low nuclear energy states remain unperturbed as the same proportion of nuclei arrive at or leave each type of site in the low energy state. If however we were to

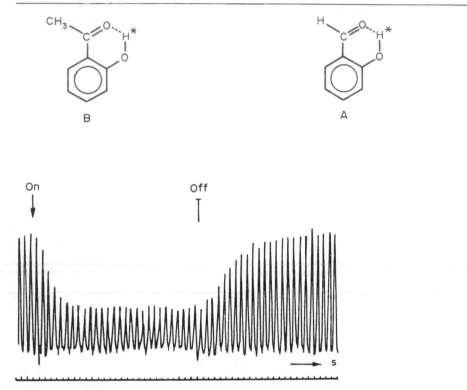

FIGURE 56

The phenolic protons of salicylaldehyde (A) and 2 hydroxyacetophenone (B) in slightly acidic solutions of a mixture give separate singlet resonances though both are broadened due to slow exchange. If the resonance of A is repetitively monitored and signal B strongly irradiated the A intensity diminishes due to transfer of equalised spin populations from B to A by exchange. The arrows indicate when the double irradiation power was switched on and off. (After Forsén and Hoffman.)

equalise the spin population at site A by saturating the resonance with high power double irradiation, then nuclei would arrive at site B with equal numbers in each state.

This would tend to equalise the nuclear population at site B and so reduce its resonance intensity. The rapidity with which the intensity falls following the application of power at site A is a function of the exchange rate while the proportionate fall in intensity is determined by the relaxation rate T_1 which operates to oppose any change. Thus both T_1 and τ can be determined from this type of experiment (Fig. 56).

100

4.5.3 *Through-space coupling via the relaxation field*

The T_1 relaxation mechanism involves an exchange of energy between a nuclear system and the degrees of freedom of motion of the whole system which gives rise to the relaxation field. Energy transfer is slow and the relaxation times of spin $\frac{1}{2}$ nuclei are long. If however we saturate one group of nuclei using a double irradiation transmitter we grossly perturb the nuclear

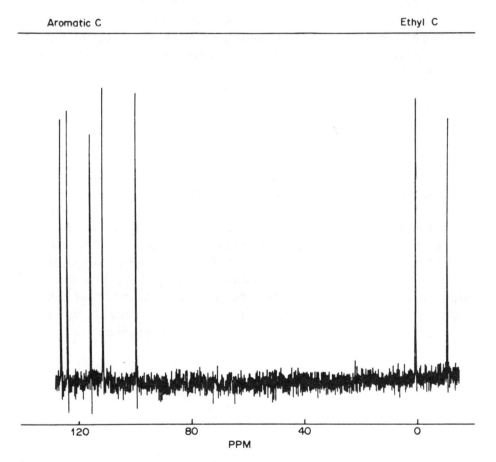

Aromatic C   Ethyl C

FIGURE 57

A ^{13}C spectrum of 3-ethyl pyridine obtained with broad band high power double irradiation of the protons so that each carbon resonance is a singlet. Resonances are observed for the seven different carbon atoms. The ^{13}C is present only to the extent of 1·1 per cent abundance, and has only 1·6 per cent the sensitivity of ^{1}H.

101

populations. The relaxation field attempts to return the population to equilibrium and in so-doing absorbs much more energy than is normal from the irradiated nuclei. If other nuclei lie close in space to the irradiated nuclei then there is a chance that they will individually provide some of this energy and undergo nuclear transitions which will alter the populations of their spin states also. A change in their resonance intensity is observed and this feature can sometimes be used to determine conformations of organic molecules. The phenomenon is known as the Overhauser effect. It occurs in all double resonance experiments and small intensity changes are often observed in the various types of experiment described previously. An example of the use of the method to determine the conformation of a compound is given later. A second important example is that ^{13}C spectra are often obtained while strongly irradiating the protons in the molecule so as to decouple them, simplify the spectra and so intensify the resonances. The Overhauser effect in this case leads to an additional increase in intensity of about three times which is very welcome. Carbons not carrying hydrogen (e.g. $C = 0$) are not affected and give rise to weaker lines. The intensity data useful for counting carbons are of course no longer reliable in such spectra (Fig. 57).

Electron spins can also produce an Overhauser effect and intensifications up to 300 times can be observed if electrons present in radicals placed in a sample solution are irradiated at their resonant frequency in the microwave region. If radicals are produced chemically in a system e.g. by irradiation with u.v. light, then their electrons, which are produced with equal numbers in each spin state, relax towards unequal spin populations and cause nuclear polarisation without the use of double irradiation, and this technique can be used to follow certain reactions. This is known as chemically induced nuclear polarisation.

The sample

5

An n.m.r. experiment involves a highly sophisticated instrument capable of resolving resonances and making measurements to a few parts in 10^9. (This is equivalent to comparing the lengths of two steel rods each 1 km long to within one thousandth part of 1 mm.) One must, therefore, accept the responsibility of preparing a sample which will not degrade the spectrometer performance. However any sample placed in the magnet gap will distort the magnetic field. Fortunately, the distortion occurs externally to a cylindrical sample and the field remains homogeneous within it except at the ends, though its magnitude is changed by an amount which depends both on the shape of the sample and upon the bulk magnetic susceptibility of the tube glass and of the sample itself. Imperfections in the glass, variations in wall thickness, variations in diameter, or curvature of the cylinder along its length all lead to degredation of the field homogeneity within the sample with consequent line broadening. For this reason high precision bore sample tubes are always used for n.m.r. Since solid particles distort the field around them, suspended solids must also be filtered from the liquid sample prior to measurement.

5.1   Standardization

We have shown that chemical shifts are invariably measured relative to a standard of some sort. There are three ways of standardizing a resonance, namely:

(i)   Internal standardization
The standard substance is dissolved in the sample solution and its resonance appears in the spectrum. This method has the advantage that the magnetic

103

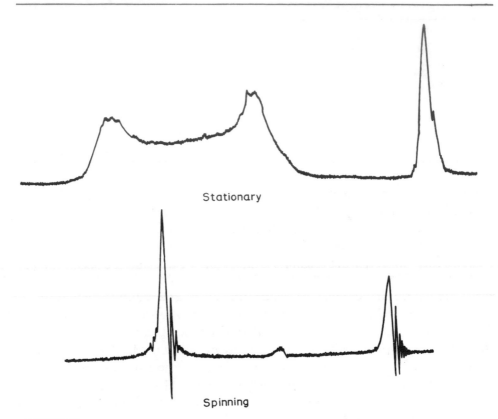

Stationary

Spinning

FIGURE 58
The field in the annular portion of a coaxial sample tube is distorted and in a stationary tube gives a broad twin-horned resonance. The capilliary resonance is the sharp one to high field of the annular signal. The distance between the horns is a measure of the volume magnetic susceptibility of the fluid in the annulus. If the tube is spun the molecules in the annulus experience an average field and a sharp singlet results. (^{1}H spectra of aqueous solutions.)

field is exactly the same at sample and standard molecules. The standard must be chosen so as not to obscure sample resonances and also must be inert to the sample. Internal standardization is the method normally used and the techniques (*ii*) and (*iii*) below are used only in special cases.

The main disadvantage of the method is that weak interactions with the solvent produce small chemical shifts which are difficult to predict and which reduce the accuracy of the measurements by an unknown amount. These shifts are said to arise from solvent effects.

(*ii*)   *External standardization*

The standard is sealed into a capilliary tube which is placed coaxially within the sample tube. The main disadvantage of the method is that since the volume magnetic susceptibilities of the sample and standard will differ by several tenths of a ppm, the magnetic fields in each will be different and a correction will have to be made for this. Since the volume susceptibilities of solutions are often not known these must be measured, so that a single shift determination becomes quite difficult if accurate work is required. The magnitude of the correction depends upon the sample shape and is zero for spherical samples so that this disadvantage can be minimised by constructing special concentric spherical sample holders. The method is used with very reactive samples or with samples where lack of contamination is important.

Because the capilliary holding the standard distorts the magnetic field around it, the field homogeneity in the annular outer part of the sample is destroyed. This can be restored by spinning, which is essential with this type of standardization. Distorted capilliaries can even then degrade the resolution and if there is any asymmetry in the annular region this will be averaged by spinning to give a field different in value from the true average, i.e. a small shift error will result. The method is nevertheless the only one suitable for measuring solvent shifts. Fig. 58 shows spectra obtained from coaxial sample tubes.

(*iii*)   *Substitution*

In this case the sample and standard are placed in separate tubes of the same size and are recorded separately in the order standard, sample, standard so that field drift between measurements can be observed and allowed for. The method is little used but is useful for weak samples of the more difficult nuclei.

5.2   Solvent effects

The solvent shift effects mentioned under internal standardization, while a nuisance to those interested simply in structure determination, are of interest in their own right, since they tell us something about the weak interactions which occur between solvent and solutes. The effect is particularly large for aromatic solvents or where specific interactions occur. The chemical shifts of a number of substances relative to TMS in a series of solvents are given below in Table 5.1 using the τ scale.

TABLE 5.1
Internal chemical shifts τ of solutes in different solvents.

Solvent:	$CDCl_3$	$(CD_3)_2SO$	Pyridine	Benzene	CF_3COOH
Solute:					
Acetone					
$(CH_3)_2CO$	7·83	7·88	8·00	8·38	7·59
Chloroform					
$CHCl_3$	2·73	1·65	1·59	3·59	2·75
Dimethyl sulphoxide					
$(CH_3)_2SO$	7·38	7·48	7·51	8·09	7·02
Cyclohexane					
C_6H_{12}	8·57	8·58	8·62	8·60	8·53

The variation in τ between solvents of course contains contributions from the solvent effect on both solute and standard. The table nevertheless is useful in indicating the existence of certain interactions involving the solutes. Thus the high field shifts obtained in the aromatic solvent benzene and for all solutes but chloroform in the aromatic solvent pyridine are obvious. Even the relatively inert cyclohexane is shifted upfield by 0·05 ppm. This arises because solutes tend to spend a larger amount of time face on to the disc shaped aromatic molecules and so on average are shifted high field by the ring current anisotropy. The magnitude of the effect depends upon molecular shape and is also increased if there is any tendency for polar groups in the molecule to interact with the aromatic π-electrons. Fig. 59 shows that the larger proportion of the space around a benzene molecule suffers ring current screening.

In the case of complex solutes each type of proton in the molecule suffers a solvent shift but because the proximity of each to solvent depends upon the shape of the molecule, each suffers a different solvent shift. For this reason a complex solute may have quite different spectra in chloroform and benzene and this fact is used to help simplify and interpret complex spectra. Fig. 60 gives an example.

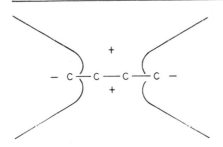

FIGURE 59
Volumes of space around a benzene ring where a proton may be screened (+) or descreened (−). There is a greater likelihood on average that the benzene will approach a solute molecule face on and therefore screen its protons. (After Johnson and Bovey.) The ring is shown edge-on.

Chloroform as a solute suffers considerable solvent shifts. The pure liquid is self-associated by hydrogen bonding but upon progressive dilution in an inert solvent the proportion of hydrogen bonded molecules is reduced and its resonance is shifted 0 29 ppm upfield. The shifts noted in the table are in excess of this and we must consider the existence of several other types of interaction. Thus specific interactions with the Lewis bases dimethyl-sulphoxide and pyridine result in low field shifts. In the case of pyridine this implies a preference for an edge on approach to the aromatic ring and therefore some ring current deshielding. In benzene on the other hand, hydrogen bonding is reduced, and there is probably face-wise interaction of the chloroform with the benzene π-electrons. Both processes tend to increase the screening so that the chloroform is shifted strongly upfield. In addition, the chloroform molecule is polar and its dipole electric field will polarize the surrounding solvent by an amount related to the solvent dielectric constant ϵ. The molecule, shown as the dipole M in Fig. 61, can be regarded as residing in a cavity whose walls become charged. This induced charge gives rise to an electric field, R, which is called the reaction field and which will also produce chemical shifts of the chloroform solute. Thus some of the variation observed in Table 5.1 will originate from differences in solvent dielectric constant.

The effect of changes in solvent dielectric constant can be clearly demonstrated by the relative solvent shifts obtained for the 4- and 2-protons of 3, 5-dimethylpyridine (3, 5-lutidine). This molecule produces an electric field as shown in Fig. 62. The field R has different components along the carbon-hydrogen bonds at the 2 and 4 positions, having its full value along

107

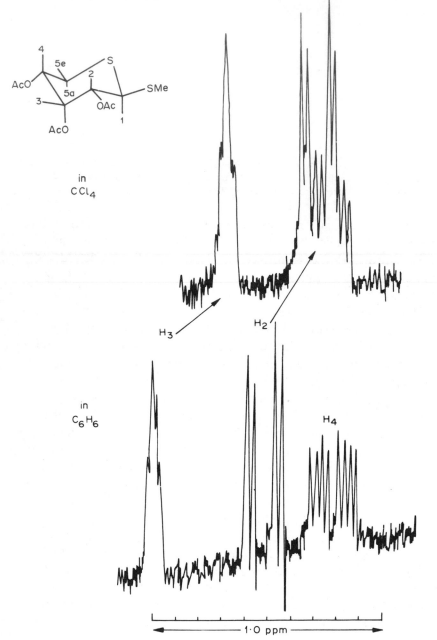

in
CCl₄

in
C₆H₆

H₃

H₂

H₄

1·0 ppm

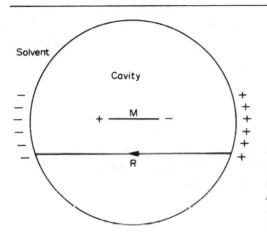

Solvent

Cavity

M

R

FIGURE 61

A polar solute molecule polarizes the surrounding solvent and generates an electric reaction field, R. The direction of R follows the motion of the solute molecule and thus can cause electric field induced chemical shifts.

FIGURE 62

The reaction field of 3, 5-dimethyl pyridine has maximum effect along the bond to the 4-hydrogen and a small and opposite effect along the bonds to the 2- and 6-hydrogens. The chemical shift between the 4- and 2- (or 6-) hydrogens therefore depends upon solvent polarizability. The small arrows near the bonds indicate the direction of electron drift.

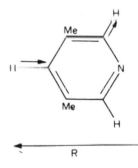

R

FIGURE 60

Part of the 90 MHz proton spectrum of a dithioglycoside dissolved in carbon tetrachloride or in benzene. The protons directly bonded to the C_5S ring are depicted by their serial number only on the inset formula. In carbon tetrachloride the 2- and 4-proton resonances overlap but become well-separated in benzene solution.

The dependence of the vicinal coupling constant upon the dihedral angle between protons is also well illustrated in this spectrum. H_2 is coupled to two protons giving a doublet of doublets. It is coupled to H_1 by 10·6 Hz. Both protons are axial, the dihedral angle is about 170° and the coupling is near the maximum (see Fig. 22). H_2 is also coupled to H_3 by 2·6 Hz. The coupling is much smaller since one proton is axial and the other equatorial to give a dihedral angle near 60°.

109

the bond to the 4-hydrogen while it is nearly zero along that to the 2-hydrogen. One would therefore expect that the 4-hydrogen would move low field as the solvent dielectric constant was increased since its electrons can be pulled along the bond away from the hydrogen, whereas the 2-hydrogen should be much less affected. Since the two atoms reside in the same fairly simple molecule one can have some confidence that specific effects and anisotropy shifts will be similar for both protons and that the observed solvent shifts between them arise primarily from the electric reaction field. The 2- (and 6-) protons resonate low field of the 4-proton. As solvents of higher dielectric constant are employed the 4-proton resonance is observed to approach that of the 2- and 6-protons so moving low field as predicted (Table 5.2).

TABLE 5.2
The effect of solvent dielectric constant upon the 2, 4 shift

Solvent	C_6H_{12}	CCl_4	$CDCl_3$	Me_2CO	MeOH
ϵ	2·02	2·24	4·81	21·3	33·62
2-4 shift ppm	1·04	0·97	0·95	0·84	0·70

Two further contributions to solvent shifts are also usually considered. One arises from the van der Waals interactions and is responsible for vapour-liquid shifts of 0·1 to 0·5 ppm. The other arises in the case of external standardization and is due to bulk diamagnetic susceptibility differences between solvents. These susceptibility shifts can be comparable in magnitude with those due to the other effects and so must be considered in any interpretation of solvent shifts, but they do not of course arise from any chemical interaction.

The various contributions to the solvent shift δ_S can be summarized by a five-term equation:

$$\delta_S = \delta_B + \delta_A + \delta_E + \delta_H + \delta_W \tag{5.1}$$

where δ_B is the bulk susceptibility contribution, δ_A is the anisotropy contribution, δ_E is the reaction field contribution, δ_H is the contribution of hydrogen bonding and specific interactions, and δ_W is the van der Waals contribution.

Instrumentation 6

6.1   Modern spectrometer systems

We have already described a basic spectrometer on page 14 and indicated how field homogeneity, field strength and stability and frequency stability must be of the highest standard. Such a spectrometer however still leaves a lot to be desired. In particular the base line is very unstable and, because of slow magnetic field drift, the spectra are not easy to calibrate. In addition this latter fault means that it is difficult to carry out a double resonance experiment since the double irradiation power keeps drifting off the irradiated resonance. These faults are overcome by two additions to the basic spectrometer.

(i) Field modulation

A pair of coils are placed on each side of the sample in the magnet gap and fed with alternating current with a frequency, depending on the instrument, in the range of 2 to 20 kHz. The main magnetic field is then no longer static but can be represented by $B_0(t) = B_0 + B_m \sin 2\pi\nu_m t$, where ν_m is the modulation frequency and B_m is the modulation amplitude. The result is that the precession frequency of the nuclear spins also varies sinusoidally in time, i.e. is frequency modulated. Under these conditions the output of the phase sensitive detector is no longer only dc but contains a sinewave oscillating at the modulation frequency ν_m. The dc component which carries the base-line variation can be blocked by a capacitor. The pure nmr signal is then obtained from the sine wave output using a second phase sensitive detector locked

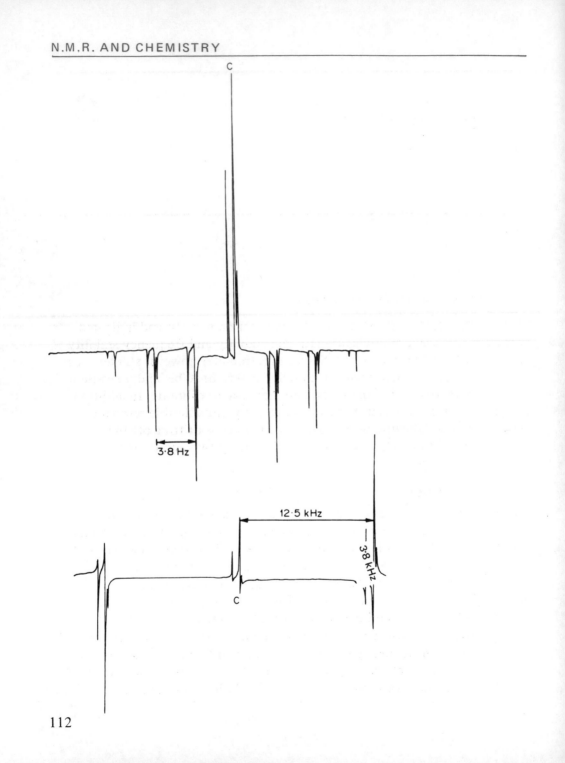

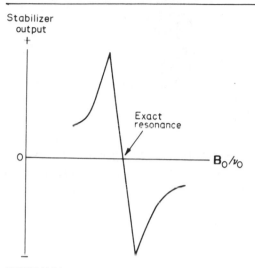

Stabilizer output

Exact resonance

B_O/ν_O

O

FIGURE 64
A resonance in the dispersion mode can be used to stabilize an nmr spectrometer. At exact resonance the output is zero but if drift occurs it becomes either positive or negative. The output can be used to provide a correction to the magnetic field which reverses the drift until the exact resonance position is regained.

by the field modulation oscillator.

The disadvantage arises that the nuclei no longer resonate only at ν_0 but also give signals called sidebands at $\nu_0 \pm n\nu_m$ where $n = 1, 2, 3 \ldots$ (Fig 63) Interference between one resonance and sidebands of another is normally avoided by using a modulation frequency high enough to separate them. The phenomenon is useful since, by altering the modulation frequency until a sideband of one resonance coincides with another (determined for instance by the shape of the wiggle beat pattern) one can accurately calibrate spectra. One can also use several different modulation frequencies simultaneously and in this way for instance produce a field lock signal (see below), a double irradiation signal, and, by using a steadily varying third modulation frequency, produce a frequency sweep. The outputs at the different audio frequencies are of course separated satisfactorily into different channels by phase sensitive detectors.

FIGURE 63
A large spectometer sweep using $3 \cdot 8$ kHz (upper) and $12 \cdot 5$ kHz (lower) modulation. The spectrum is the proton resonance of acetaldehyde, $CH_3 CHO$, containing TMS (spin coupling not resolved) and its three bands are repeated in each sideband. The relative intensities of the centreband (c) and sidebands are determined by the current in the modulation coils (modulation amplitude) and the phase settings of the rf and af phase detectors. The af phase settings were different for the two traces; neither was adjusted ideally.

(ii) Field-frequency locking

In order to obtain maximum stability it is necessary to stabilize both field and frequency. A convenient single device which will detect changes in both is based on the nuclear resonance of the standard signal in our sample. The spectrometer channels just described are quite independent. If we adjust the phase in one channel to give a dispersion signal of the standard resonance, we can use this to determine whenever the spectrometer drifts off the resonance condition. At resonance the output is zero, but if drift occurs either a positive or a negative output is obtained, the sign depending upon the direction of drift, and we can use this output to provide a correction voltage to alter the magnetic field until the output is again zero (Fig.64). By using this technique a modern spectrometer can stay on resonance (i.e. keep the ratio B_0/ν_0 constant) indefinitely. A second channel operating in the absorption mode is frequency swept (e.g. by altering ν_m) to develop the spectrum. Calibration of the frequency sweep is done easily using an electronic counter to measure the difference in frequency between ν_m in the two channels.

6.2   Noise elimination

It is an unfortunate fact that n.m.r. is not a sensitive technique. We rely for our signal on a very small difference between upper and lower state nuclear populations and even then are hampered by the saturation effect, so that random electronic noise often obscures the weaker resonances. For this reason there has been, and is, considerable preoccupation with ways of increasing the signal-to-noise ratio of n.m.r. spectrometers. Improvements in electronic devices mean of course that modern spectrometers considerably exceed the performance of their predecessors. Much attention has however been given to different ways of improving performance even further. A few words on how this has been done are included here, since the principles apply in general to any form of spectrometry.

(i) Filtration

This is the name given to the operation of separating wanted and unwanted signals and is analogous to the chemical and optical procedure of the same name, i.e. we need a device which will react differently to the two sets of signals.

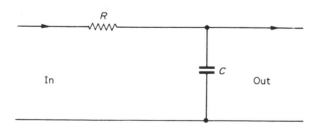

FIGURE 65
The spectrometer output consists of wanted signal plus random noise. If the signal is less in amplitude than the noise it cannot be seen.

The output of an n.m.r spectrometer consists of random noise at all frequencies and a slowly varying dc voltage which is the signal (Fig. 65). These can be separated using the circuit shown in Fig. 66. The capacitor offers low impedance for all the noise frequencies except the lowest and considerably tidies up the signal. Unfortunately it also distorts the signal. This can be seen by reference to Fig. 67. We see that the signal is delayed, it is diminished in intensity and the lines are broadened, so reducing the resolution. For acceptable performance the time constant RC must be about 20 per cent of the time taken to traverse a line. The noise level is proportional to $1/\sqrt{(RC)}$ so that to reduce the noise by 10 times we have to increase RC by 100 times from its usual value and sweep 100 times more slowly. If the time is available such scans can give good results, but of course they demand high spectrometer stability.

FIGURE 66
A simple resistance capacitance filter network.

115

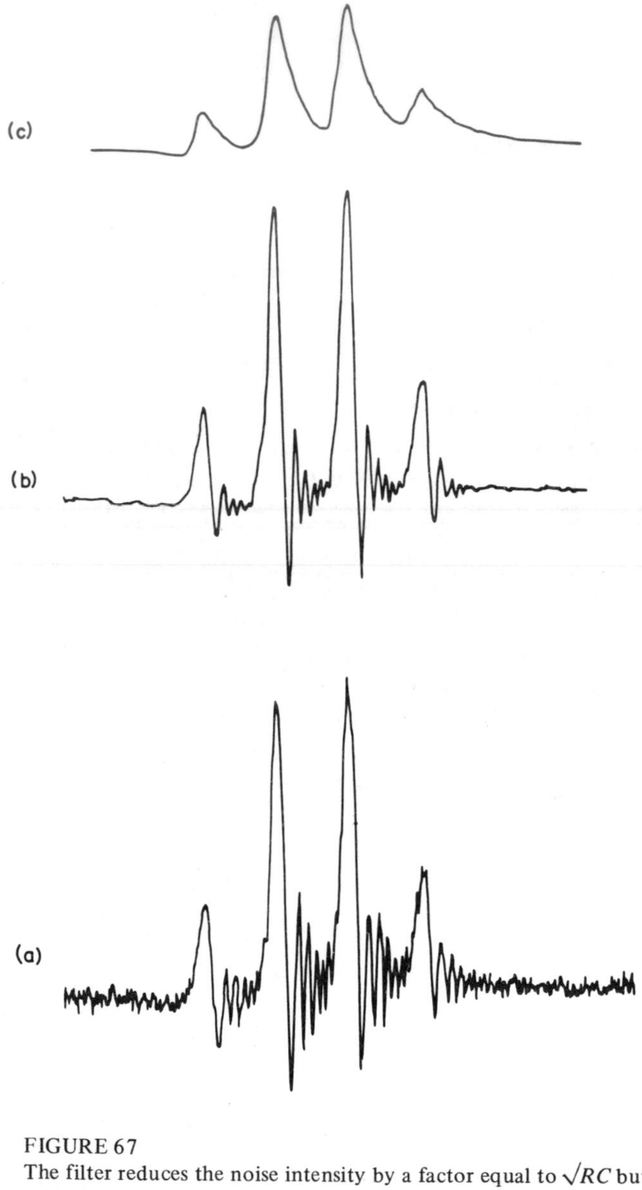

FIGURE 67
The filter reduces the noise intensity by a factor equal to $\sqrt{RC}$ but also distorts the signal. The three quartets shown were recorded under identical conditions except that the filter time constant RC, was altered: (a) $RC = 0.01$ sec; (b) $RC = 0.1$ sec; (c) $RC = 1.0$ sec. In (c) the noise is very much reduced (by ten times) but the signal is also distorted and its intensity reduced by half, its resolution is much poorer and it is displaced 0.5 Hz upfield due to the retarding effect of the filter. The spectrum was drawn from left to right.

116

(ii) CAT-ing or spectral accumulation

CAT is essentially an n.m.r. jargon term derived from the acronym for computer averaging of transients. To cat one needs a multi-channel analyser. This is a device with many (e.g. 500,1000 or 4000) memory boxes or channels for storing binary numbers. The output of the spectrometer is sampled at regular intervals and converted by an analogue-to-digital converter into numbers proportional to the voltage, and the numbers obtained during a sweep placed sequentially in the analyser channels. Thus the cat now contains the spectrum. If we arrange it that the device which selects the channels also drives the frequency sweep (or vice versa) then if a second sweep is made each individual channel will always be fed with a number corresponding to the same part of the spectrum. We can thus sweep the spectrum several times and add spectra together. The numbers in those channels corresponding to signals build up proportionally to the number of sweeps N_s, whereas the noise, which is incoherent and is sometimes positive and sometimes negative, only increases as $\sqrt{N_s}$. One hundred sweeps therefore gives a ten times increase in signal-to-noise ratio. The time taken is thus the same as for the previous technique. However the method does give some improvement because the faster sweeps reduce the likelihood of saturation and allow more transmitter power to be used.

Since the term cat was introduced, the term dog has been coined for slow sweeps with large time constant (method i).

(iii) Fourier transform spectrometry

If we irradiate our sample with a pulse of rf power of width t seconds then this is equivalent to using irradiation of bandwidth of $1/t$ Hz. If t is short enough the bandwidth is large and we stimulate all the nuclei into giving a signal simultaneously. The resultant transient signal, which follows the pulse, contains frequencies corresponding to all the frequency separations in the spectrum. If the transient output is collected in a time-swept cat memory the numbers in the channels can be fed to a small computer and a Fourier Transform calculated. This operation produces the normal analogue spectrum. The method has the advantages that a single transient response can be collected in a very short time of between 0·4 and 3·0 seconds (the longer the time the better the resolution), and that several thousand responses can be accumulated to give an excellent signal-to-noise ratio in a reasonably short time. The time

117

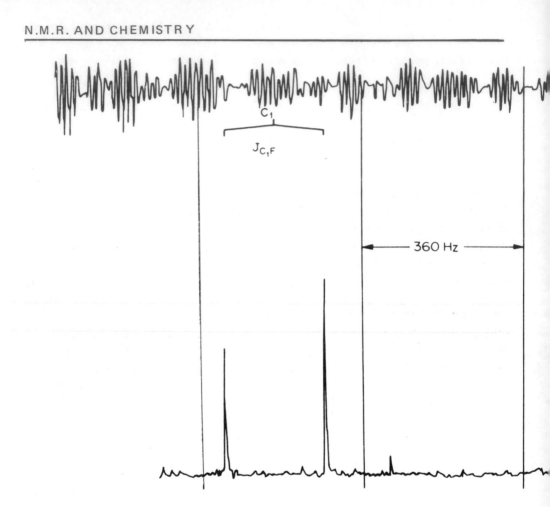

FIGURE 68

^{13}C transient response (upper trace) obtained from fluorobenzene after 500 pulses applied in 205 seconds. The protons are strongly decoupled using broad band double irradiation. The lower trace is the computed fourier transform. Observe the reduced noise level when compared with Fig. 57. It is possible to observe spin spin coupling between the fluorine atom and all four types of carbon.

necessary is the same whether a wide chemical shift range is covered or a narrow one. Since if we cat normally we have to sweep more slowly the

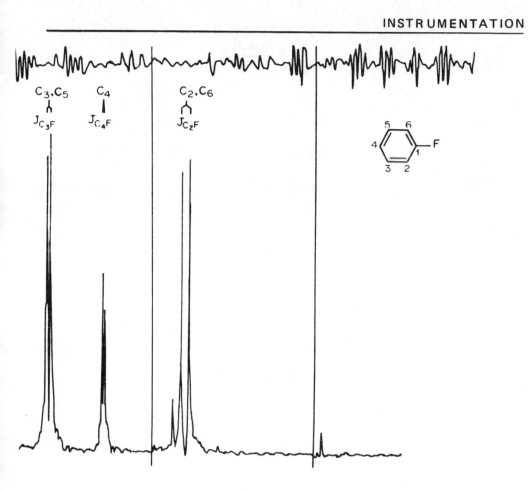

wider the spectrum covered, so as not to traverse the lines too rapidly, Fourier transform spectroscopy reduces the time needed to obtain a spectrum by a factor equal to spectral width/line-width. This can be as high as 1000 times in some cases. Fourier transform spectroscopy is thus useful where signals are very weak and widely spread out and has come to be much used for ^{13}C spectroscopy, especially in association with high power proton double irradiation (Fig. 68).

119

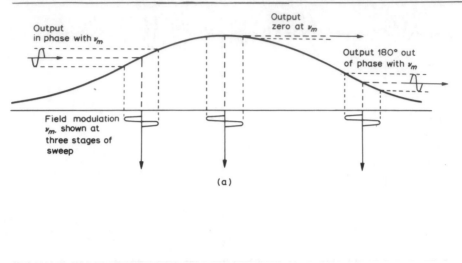

(a)

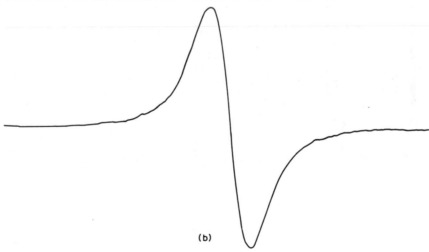

(b)

FIGURE 69

(a) Production of a derivative spectrum. The field is modulated sinusoidally at a low frequency and with a sweep width much smaller than the line-width. A sinusoidal output is obtained whose amplitude depends upon the slope of the absorption curve of the resonance. The phase of the output changes 180° on passing the maximum and so the sign of the output reverses. (b) A derivative spectrum of ^{45}Sc (aqueous salt solution). The peak to peak width is 700 Hz.

120

6.3   Derivative spectroscopy

This mode of operation is often used where the broad lines of solid materials are being observed. Field modulation is used but at such a low frequency that the sidebands occur within a line-width. Typically frequencies around 35 Hz are chosen. A small modulation amplitude is used and this results in effectively a repetitive sinusoidal field sweep of a small portion of the resonance. The receiver output varies sinusoidally at the same frequency and the magnitude of the output depends upon the difference between resonance height at the start and end of the sinusoidal sweep, i.e. it is proportional to the slope of the signal envelope (fig. 69). The line is similar in appearance to the dispersion signal but its detailed shape is quite different. The method is excellent for very broad lines where a large modulation amplitude can be used, but it is not useful for mixed broad and narrow lines since the large modulation amplitude needed to observe the broad component also broadens the narrow component. In this case dispersion working with a cat seems to be the current solution.

Part II

The uses of n.m.r. in chemistry   7

This chapter is divided into a series of sections each one dealing with a particular topic, for example with structural determination of organic molecules or the study of dynamic systems. The object of this chapter is to indicate how the various principles outlined in the previous chapters are applied in practice. Some of the examples are presented as problems while the more complex ones are wholly or partly worked through.

An attempt has been made to cover as wide a variety of topics as is possible. This means unfortunately that no single topic can be given a very full treatment and particularly that the use of nmr by the preparative chemist for the determination of the structures of organic compounds is given a far smaller proportion of space than its use deserves. This is inevitable in a book of this size and the student is referred to the bibliography where he will find mention of several books which cover this aspect more fully.

7.1   The determination of the structures of organic molecules using proton spectra

One example will be worked through and the remainder are left for the student to attempt to solve. The formulae of the substances giving the spectra are reproduced at the end of the section but are not assigned to individual spectra. That is left to the student.

Several complementary pieces of information can be obtained from an nmr spectrum and should be considered in turn when analysing a trace. First we have the chemical shift information and a guide to the meaning of this can be found from the simplified chart of Fig. 70. Secondly there is the spin

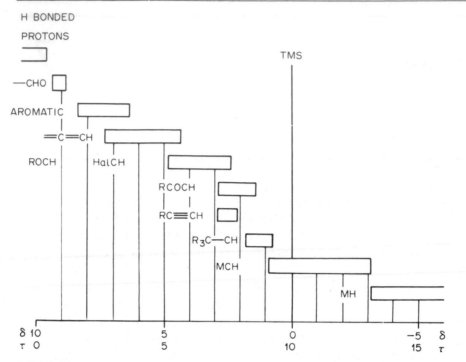

FIGURE 70
Chart of approximate chemical shift ranges of protons in organic and organo-metallic compounds. M represents a metal and Hal a halogen.

coupling information which allows us to count numbers of proximate protons. Thirdly there is the intensity data which allows us to determine the relative numbers of each type of proton. This is not necessarily the same thing as is found from spin coupling data, since for instance a compound might contain one, two, or more identical ethyl groups. Intensities can sometimes be obtained by inspection, but because lines have different breadths the area is a better guide. For this reason spectrometers are equipped with an integrating device. The numbers of protons determined by integration are indicated by the numbers on the spectra.

Consider the spectrum (Fig. 71) of α-chloro propionic acid. This contains three groups of resonances, one at low field, one at 4.45 ppm, and one at 1·75 ppm. The high field doublet is split by spin coupling since the same splitting is observed in the quartet. Thus we have three sorts of proton.

126

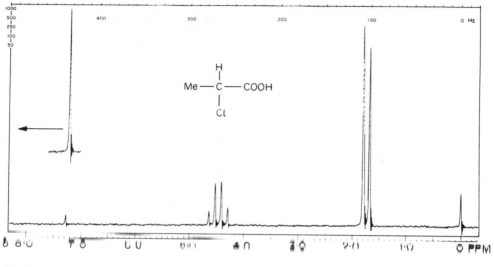

FIGURE 71
60MHz proton spectrum of α-chloropropionic acid.

Inspection of Fig. 70 suggests that these belong to the fragments HC-C, HC(C=O)R, and a hydrogen-bonded proton. The doublet-quartet AX_3 part of the spectrum has a typical vicinal coupling constant of 7 Hz so that we have the fragment CH_3-CH(CO)R. The intensities suggest proton ratios of 3 : 1 : 1. If we were given that the empirical formula is $C_3H_5O_2Cl$ then we could write the structural formula as CH_3-CH(CO)OH and place the chlorine atom appropriately to satisfy the valence rules.

Now apply the same approach to Figs. 72 to 79. The formulae are grouped in Fig. 80 with one extra inserted to reduce the possibility of analysis by elimination. Bear in mind: (*i*) that the fluorine resonances of the fluorine-containing compounds are not visible, though splititng due to fluorine may be; (*ii*) that broadened NH proton resonances may nevertheless cause splitting in vicinal protons; (*iii*) that the outer lines of a multiplet with many lines may not be visible (use Pascals triangle to check the multiplicity of the apparent quintet of Fig. 78); and (*iv*) that alcoholic protons usually exchange and are variable in position.

127

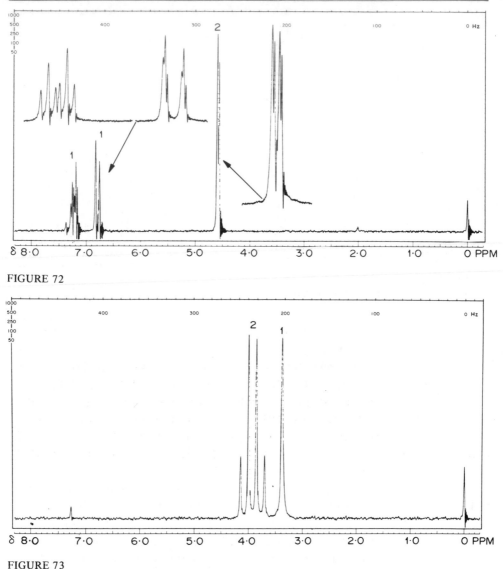

FIGURE 72

FIGURE 73

FIGURE 72–80

60MHz proton spectra of eight of the substances depicted in Fig. 80. The chemical shifts are on the δ scale. Where resonances appear below 8 ppm they have been recorded with the spectrometer scale offset, to bring the resonance onto the paper.

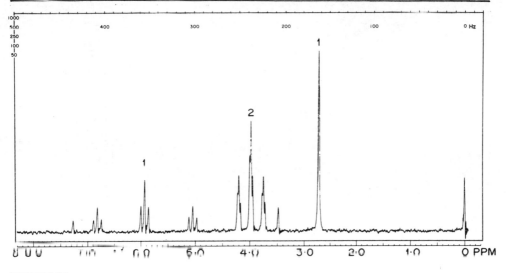

FIGURE 74

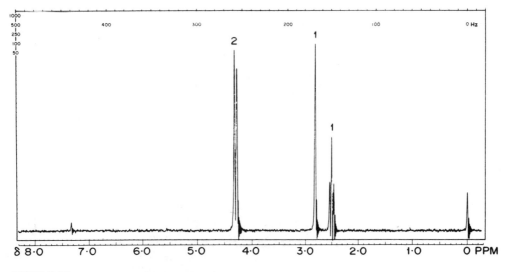

FIGURE 75

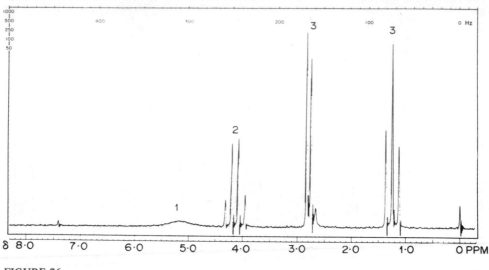

FIGURE 76

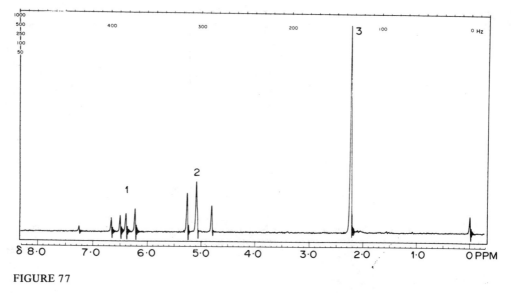

FIGURE 77

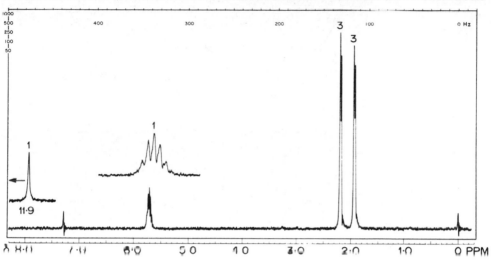

FIGURE 78

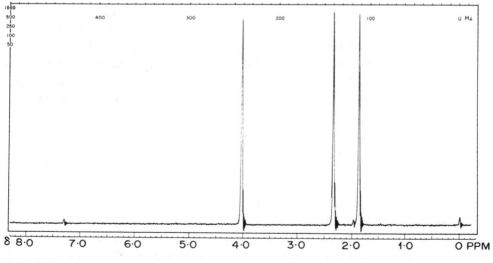

FIGURE 79

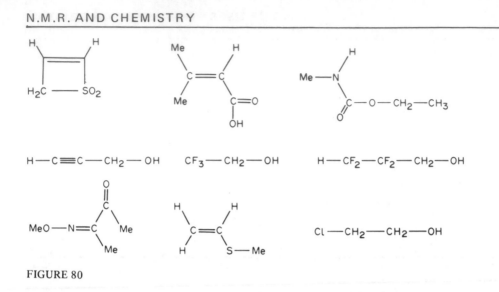

FIGURE 80

7.2 Determination of the conformations of organic molecules

Cyclohexane is known to exist in the chair configuration, of which there are two forms, a given proton being equatorial in one and axial in the other (Fig. 81). Interconversion of the two forms is rapid at room temperature and so the protons experience an average environment and the proton resonance is a sharp singlet. If the sample is cooled the interconversion becomes slower and the axial and equatorial environments become distinguishable, with the result that the resonance broadens and separates at a low enough temperature into two resonances. Since the equatorial-axial geminal pairs of protons are spin coupled the two resonances are split into doublets and can be regarded crudely as an AB quartet. The axial-equatorial chemical shift is quite large, of the order of 0·5 ppm (Fig. 82).

In the case of the six-membered cyclic molecules of the glycosides the rate of ring flexing is often reduced by the presence of a bulky substituent, which tends to take an equatorial position because of the steric crowding which occurs in the axial position. If the hydroxyl groups are acetylated the OH resonances are eliminated, and clear spectra of the protons on the ring skeleton are obtained. Their chemical shifts extend over a large range and the resulting spin coupling patterns between them are first order and can be used to determine the conformation of the molecule. This is possible be-

132

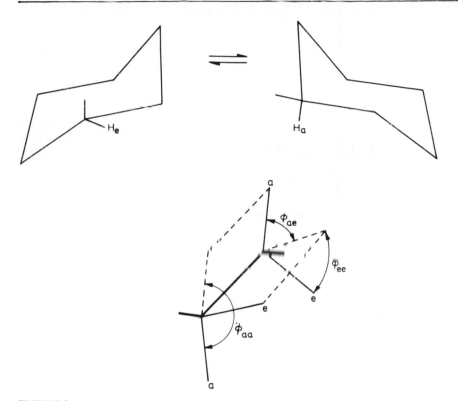

FIGURE 81

Upper: Interconverting chair forms of cyclohexane, C_6H_{12} indicating how one proton changes from an axial (a) to an equatorial (e) position. Lower: A view of a carbon-carbon bond of cyclohexane showing the various dihedral angles between the carbon-hydrogen bonds. The hydrogen positions are designated a or e.

cause the dihedral angle between pairs of vicinal protons is large (near $180°$) in the case of two axial protons but much smaller and near to $60°$ in the case of two equatorial or an equatorial and an axial proton (Fig 81). Reference to the Karplus curves (Fig. 14) will show that we would expect the spin-spin coupling between two axial protons to be much larger than that between two equatorial protons or between an equatorial and an axial proton.

Using this rule we can obtain the configuration of the dithioglycoside whose formula and spectrum are shown in Fig. 83. The spectrum can be assigned to individual protons as follows. The proton H_A is the only one

133

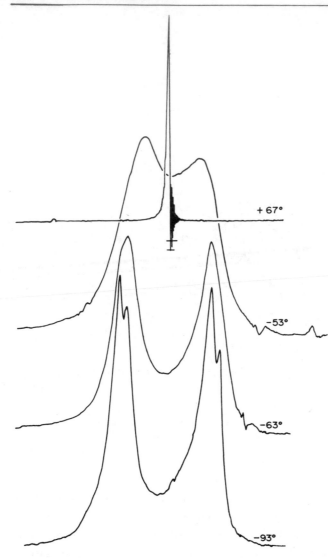

FIGURE 82
Proton spectra of cyclohexane over a range of temperatures. At a low enough temperature separate resonances can be observed for axial and equatorial protons.

134

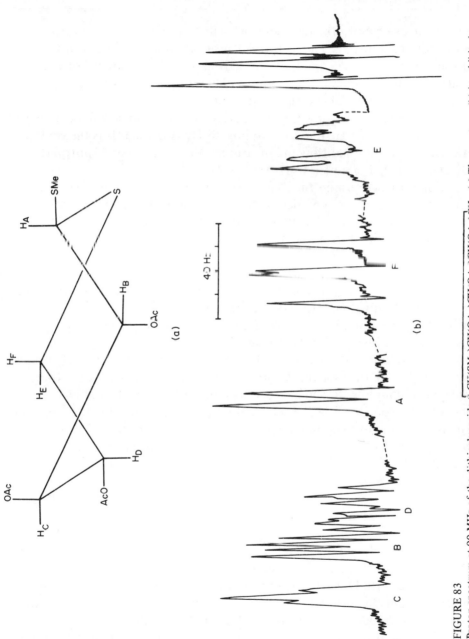

FIGURE 83

Proton spectrum at 90 MHz of the dithioglycoside S CH(SMe)CH(OAc)CH.OAc)CH(OAc)CH$_2$. (a) The spectrum (b) is exhibited as a series of expanded sections to show up the fine structure. The full spectrum extends over 3·7 ppm. The three high-field lines due to the four methyl groups in the molecule are recorded at one-tenth the receiver gain coupled to a single proton and therefore will be the only one exhibiting simple doublet splitting. Thus multiplet A arises from H_A. The coupling constant of 5 Hz is small and must arise from an axial-equatorial or equatorial-equatorial interaction. Since the bulky S Me group will take up an equatorial position then H_A is axial and H_B is equatorial.

H_B is coupled to two protons and therefore its resonance will be a quartet with one spacing of 5 Hz due to coupling to H_A. Multiplet B is the only one which satisfies this condition and measurement of the second splitting indicates that H_B is coupled to H_C by 2·8 Hz.

Looking further around the ring we see that H_D is also a unique proton in that it is the only one coupled to three other protons. Its resonance should therefore consist of a doublet, of doublets, of doublets i.e. eight lines, all of equal height. Multiplet D is the only one satisfying this condition. Measurement gives three coupling constants of 11·4, 4·0, and 2·8 Hz, the first being a large, axial-axial coupling which is repeated only in multiplet F. Multiplet F is a quartet with a second splitting of 13·1 Hz which is repeated in multiplet E. Now two of our protons, H_E and H_F, are a geminal pair and a large coupling constant is expected. Multiplets D, E, and F are the only ones exibiting large coupling constants but D is assigned to H_D which has no geminal proton. Therefore multiplets E and F arise from the geminal pair. H_D and the proton giving rise to resonance F must both be axial to account for the large vicinal coupling. Resonance F thus must arise from H_F and by elimination resonance E is due to H_E. Measurement of multiplet E indicates that H_E is coupled to H_D by 4·0 Hz, a small value as expected. This leaves one coupling of 2·8 Hz to H_D unaccounted for and this must arise from coupling to H_C. It is a small value and H_C must be equatorial. H_C is coupled to H_B also by 2·8 Hz so that the two centre lines of its quartet overlap and it is observed as a triplet; multiplet C. The coupling constant between H_B and H_C is small, consistent with both being equatorial.

Another example where conformational information was derived from an nmr spectrum is shown in Fig. 84. The spectrum contains an ABXM multiplet in which the X part of an ABX spectrum is further coupled to the NH proton (i.e. strictly an ABMX system). The values of $J(A-X)$ and $J(B-X)$ can be calculated from the multiplets at 3·0 and 4·3 ppm and this gives a measure of the dihedral angles between H_X and H_A or H_B, which are found to be highly dependent upon the solvent used; the ester substituent at position 3 being equatorial in chloroform and axial in pyridine.

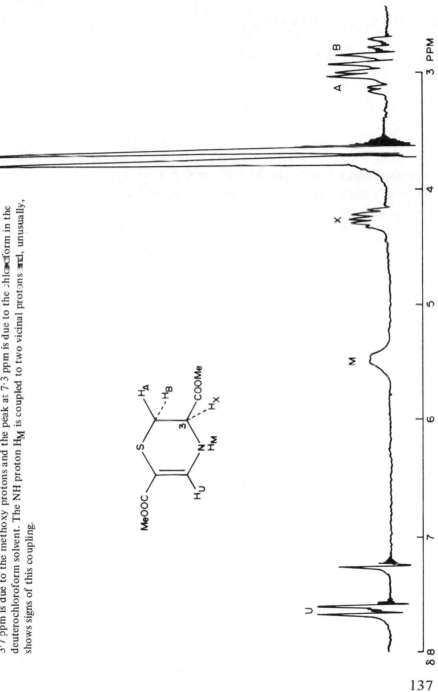

FIGURE 84

90 MHz proton spectrum of 3, 6-dimethoxycarbonyldihydro-1, 4-thiazine, a substance related to penicillin. The subscripts indicate the assignment of the resonances. The intense doublet near 3·7 ppm is due to the methoxy protons and the peak at 7·3 ppm is due to the chloroform in the deuterochloroform solvent. The NH proton H_M is coupled to two vicinal protons and, unusually, shows signs of this coupling.

137

7.3   Two double resonance studies of organic compounds

7.3.1   As an aid in solving a complex structure

A degredation product from the macrolide antibiotic tylosin was a C_6-dihydroxy acid which was converted to a C_{11}-diacetyl methyl ester, $C_{11}H_{16}O_6$. The proton spectrum of this compound is shown in Fig. 85. The sharp singlets at δ 3·85 and δ 2·17 ppm arise from the ester and acetyl methyl groups respectively. The quartets at δ = 3·0 and 5·12 ppm are each due to a single proton and the doublet at δ = 4·27 ppm is due to two protons. The pattern high field of δ = 2·0 ppm is due to five protons and has the appearance of an ethyl group with extra coupling to the methylene protons. The quartet splitting of the single protons does not arise from coupling to the methyl groups, since these are singlets, and in any case are too distant to couple appreciably. There are no other groups of three nuclei and the

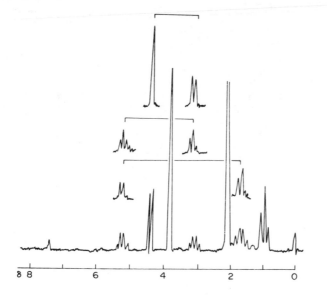

FIGURE 85

60 MHz proton spectrum of a diacetyl methyl ester of a degredation product of tylosin. The empirical formula is $C_{11}H_{16}O_6$. Three field-sweep double-resonance experiments were carried out and the simplified form of the multiplets obtained are shown above the main trace. The horizontal brackets show which pairs of multiplets were decoupled in each experiment. (After Morin and Gorman.)

quartet splitting must arise because of equal coupling to two or three groups of inequivalent protons containing a total number of three protons. All the resolved couplings have a value indicating that the coupling arises between vicinal protons.

The above data are sufficient to enable the structure of the compound to be guessed. Double resonance studies make possible a much more definate conclusion. Three pairs of experiments were carried out using a field sweep technique. The main field is modulated at a frequency equal to the separation of a pair of multiplets. This produces a centreband signal and its sidebands. Thus when a sideband coincides with one multiplet the centreband coincides with the other. If the centreband were used for irradiation and the modulation adjusted to give low power in the sidebands, then when either ol the multiplets in observed using a suitable sideband the other is being decoupled by the centreband. It the multiplets observer are spin coupled then the spectra traced out will be simplified. The upper short traces in the figure show the multiplet shapes obtained for three such experiments. These show which multiplets are coupled and how many protons are involved in the interaction. Thus, for the uppermost pair, irradiation of the δ 4·27 doublet reduces the coupled δ 3·0 quartet to a doublet indicating that the quartet splitting arises from coupling equally to a single proton and to a proton pair.

The only possible structure consistent with these results is:

$$CH_3 - CH_2 - \overset{\overset{\displaystyle H}{\displaystyle |}}{\underset{\underset{\displaystyle AcO}{\displaystyle |}}{C}} - \overset{\overset{\displaystyle H}{\displaystyle |}}{\underset{\underset{\displaystyle COOCH_3}{\displaystyle |}}{C}} - CH_2 - OAc$$

7.3.2 *As an aid in determining a conformation*

The aziridine, 3-carbomethoxy-5, 5-dimethyl-4-thia-1-azabicyclo [4.1.0] hept-2-ene (Fig. 86a) can exist in two possible conformations which are shown in Fig. 86b, A and B. Conformation A was expected to be the preferred one, since in B the α-methyl group eclipses the C_6-C_7 carbon-carbon bond. The 90 MHz proton spectrum of the compound in $CDCl_3$ is shown in Fig. 86c. This consists of a singlet at $\delta = 7\cdot76$ ppm for the vinylic 2-proton, three intense singlets at $\delta = 3\cdot7$, $1\cdot61$, and $1\cdot02$ ppm for the methyl protons of the ester and the *gem* dimethyls respectively and three small

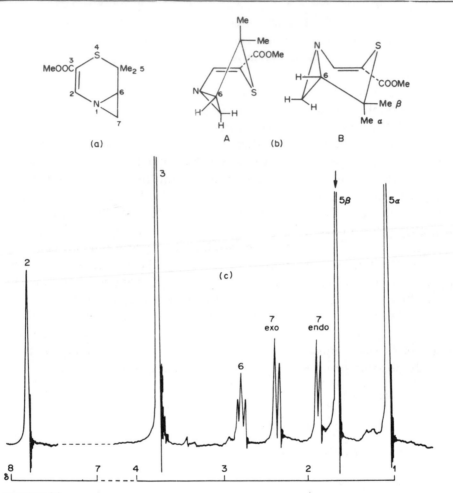

FIGURE 86

(a) Structural formula of the aziridine 3-carboxymethyl-5, 5¹-dimethyl-4-thia-1-aza-bicyclo [4.1.0] hept-2-ene. The ring positions are numbered 1 to 7. (b) The two possible conformations of the aziridine. The 6-hydrogen lies between the gem dimethyl groups in A but is eclipsed by the β methyl in B. (c) The 90 MHz proton spectrum of the aziridine. The resonances are numbered corresponding to the associated ring positions. Note that the δ scale is broken between 4 and 7 ppm to allow the spectrum to be expanded. (d) Representative integral traces over the 6 and 7-*exo* resonances obtained normally (lower trace) and with double irradiation of the 5β-methyl resonance (upper trace) demonstrate the increase in integral obtained due to the Overhauser effect. The horizontal scale is expanded relative to (c). The vertical height of each step of the trace is proportional to the area of the resonance which would be traced out in the normal absorption mode spectrum.

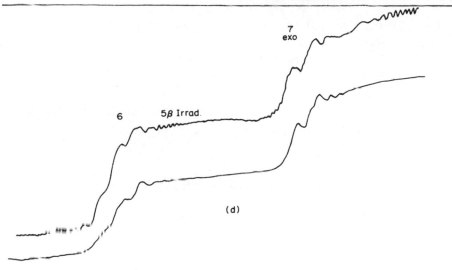

FIGURE 86 (d)

multiplets for the three protons (6, 7, 7′) associated with the three-membered
ring. The low field one arises from the 6-proton and is coupled to the *endo*
and *exo* 7-protons thus appearing as a triplet (it is a quartet on a more
expanded scale) while the 7-protons are not resolvably coupled to each other
(a case where J *gem* $\approx$ 0) and are observed as doublets. Comparison of the
5-methyl shifts with those in analogous compounds of known conformation
suggested that surprisingly the conformation was B with 6-H close to the
β-methyl and the *endo* 7-H close to the α-methyl. In order to obtain further
evidence an internuclear Overhauser experiment was carried out. The
spectrometer field was locked to the TMS signal and frequency sweep spectra
were obtained for the 6-H, 7-H region of the spectrum in the integrated mode.
Integrals were obtained without double irradiation and with sufficiently
strong double irradiation of either of the 5-methyl peaks to saturate them. It
was found that irradiation of the highest field methyl peak lead to a ~ 45 per
cent increase in integrated intensity of the *endo* hydrogen at position 7 while
irradiation of the low field methyl peak (arrowed in the figure) lead to a
~ 35 per cent increase in the intensity of the peak due to the hydrogen at
position 6 (Fig. 86d). This clearly points to conformation B since in the
case of conformation A the 6-hydrogen lies between the α and β 5-methyl
groups and its resonance should be intensified by the irradiation of either

141

methyl peak, though since the internuclear distances are larger in A than in B then the effect should be weaker.

Integration of the resonances was carried out several times and the results averaged since the intensity changes are quite small. Integration is essential since marked changes in the height of the 6, 7 resonances occur upon irradiation of the 5-methyls, due to the elimination of small, unresolvable, long-distance spin coupling interactions.

7.4 The determination of the structures of transition metal complexes

A vast array of transition metal complexes is now known in which the central metal atom is bonded to several of a possible multitude of monodentate ligands, e.g. carbonyl, CO; hydride, H; a phosphine, PRR'R''; a halide; or an organic group. Complexes are also formed with bidentate ligands such as ethylenediamine and with unsaturated organic systems *via* the π-bonds. The stereochemistry of such complexes offers considerable possibility of variation and nmr has been found to be invaluable in deciding just how the ligands are arranged around the metal atom. The nmr spectrum of a ligand is usually that expected to arise from its structure though the chemical shifts of the resonances of the free and complexed ligand and the coupling constants will be different. There are however a number of other features which are of use to the chemist, for instance the very high field shift of hydride protons is diagnostic of their presence and spin coupling may be observed between ligands or between ligand and metal if the metal has a magnetically active isotope. Several other examples follow.

Two arrangements are possible of the methyl and ammine ligands in the octahedral platinum complex $Pt(CH_3)_3 (NH_3)_3$, namely with the methyls (or ammines) all mutually *cis*, the *all cis* configuration, or else with two of the methyls *trans*, the *trans meridional* configuration (Fig. 87a and b). There is only one sort of methyl group in the *all cis* complex and so the three methyls are isochronous. The expected singlet is however split into a $1:4:1$ triplet due to coupling to the platinum isotope, ^{195}Pt, for which $I = \frac{1}{2}$ and which is 33·7 per cent abundant.

$$^2J\,(^1H - \,^{195}Pt) = 75\,Hz$$

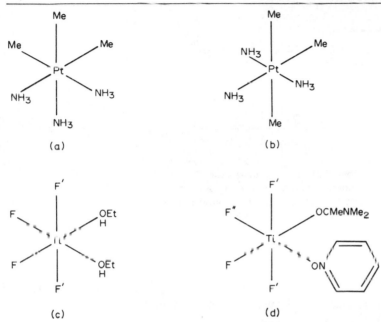

FIGURE 87

(a) All *cis* configuration of Pt(Me)$_3$(NH$_3$)$_3$; (b) *trans meridional* configuration of Pt(Me)$_3$(NH$_3$)$_3$; (c) and (d) tetrafluorotitanate complexes with two *cis* ligands. The different types of fluorine atom are indicated by primes.

In the *trans meridional* complex however one methyl is *trans* to an ammine while the other two are *trans* to methyl. They are no longer all in the same chemical environment and two resonances result with the intensity ratio 1 : 2. Each one is of course also split into a 1 : 4 : 1 triplet but now with different values of $^2J(^1H - {}^{195}Pt)$.

One should note the large value of $^2J(^1H - {}^{195}Pt)$ in the above example. Coupling to directly bonded protons is even larger; in planar (Et$_3$P)$_2$Pt HCl, $^1J(^1H - {}^{195}Pt)$ has the value 1276 Hz (Fig. 88).

A more complex spectrum is observed in the case of the octahedral tetrafluorotitanate complexes TiF$_4$L$_2$ (L = a ligand) depicted in Fig. 87 (c and d). A singlet fluorine resonance is observed in organic solvents at room temperature due to rapid interchange of fluorine positions, but at low temperature the structure is 'frozen' sufficiently long for the full nmr spect-

143

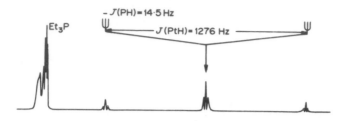

FIGURE 88

60 MHz proton spectrum of *trans* Pt HCl(PEt$_3$)$_2$ in benzene. The ethyl resonance of the phosphine ligands appear at low field while the hydride resonance is 16·9 ppm high field of TMS. The hydride resonance is split into a triplet by coupling to the two equivalent phosphorus atoms, ^2J(P-H) = 14·5 Hz and has identical platinum satellites due to coupling to the 33·7 per cent of ^{195}Pt present. (After Chatt and Shaw.)

rum to emerge. In the case of the di-ethanol complex (c), we observe two resonances separated by 60 ppm due to fluorine type F and fluorine type F′. The pairs of fluorine atoms are however spin coupled and give rise to an A$_2$X$_2$ spectrum, i.e. to two triplets with J(F-F′) = 36 Hz. Such a spectrum can only arise if the ethanol ligands are mutually *cis*. If the two organic ligands are different (d) then an even more complex spectrum results since the fluorine atoms *trans* to the ligands are no longer isochronous. The F′ fluorines resonate close to their position in complex (c) whereas the F fluorine is 42·3 ppm and the F″ fluorine is 28·9 ppm down field. These shifts are sufficient for a first order A$_2$MX spectrum to be obtained. The two equivalent F′ fluorines split both F and F″ into triplets (J(F-F′) = 39 Hz and J(F″-F′) = 35 Hz) while F and F″ interact mutually and each exhibits a 1 : 1 doublet splitting (J(F-F″) = 48 Hz). The fluorine spectrum thus consists of two doublets of triplets each of intensity one, and a doublet of doublets of intensity two due to fluorines F′.

The student should attempt to reconstruct this spectrum from the information given using squared paper.

This example also shows the very large effect on nmr parameters which is exerted by *trans* ligands. Thus the fluorine *trans* to ligands in (d) are 17·7 and 31·1 ppm up field of those *trans* to ligands in (c). In contrast the *cis* fluorines are hardly affected by the change and nmr forms a convenient tool for studying *cis/trans* effects in complexes.

The study of the methyl proton resonances of tertiary methyl phosphines in complexes containing several such ligands has also provided a very useful guide to their stereochemistry. Pairs of methyl phosphines can be considered as forming the four spin system

H . . . P . . . (Metal) . . . P' . . . H'

in which for simplicity only one proton is considered per phosphine and the carbon atoms are omitted. All nuclei have $I = \frac{1}{2}$. The primed and unprimed atoms are not magnetically equivalent since $J(P-H)$ and $J(P-H')$ are not the same. Since both P and P' and H and H' have the same chemical shifts a second-order spectrum results. This can be simply analysed by dividing the full spectrum into subspectra as follows:

(i) When both phosphorus spins are +. The two hydrogen atoms are isochronous and give rise to a singlet.

(ii) Similarly when both phosphorus spins are − another singlet is obtained. The separation between the two singlets is $|J(H-P) + J(H-P')|$, its magnitude depending upon the relative signs of the two couplings. Thus the spectrum always contains a 1 : 1 doublet.

(iii) When the two phosphorus atoms have different (opposite) spins the two protons are no longer isochronous and can exibit spin coupling. However $J(H-H')$ is usually negligibly small but virtual coupling occurs via the strong H-P and P-P' interactions such that an AB spectrum results with each half based on one of the singlets and with an apparent coupling equal to $J(P-P')$ (Fig. 89a). If $J(P-P')$ is small, i.e. $J(P-P') \ll |J(H-P) + J(H-P')|$ then the lines of the AB sub-spectrum are close to the singlets and may not be resolved, so that a deceptively simple doublet results (Fig. 89b). Alternatively if $J(P-P')$ is large so that $J(P-P') > |J(H-P) + J(H-P')|$ then a closely coupled AB sub-spectrum is obtained with the two central lines close together and containing virtually all the sub-spectral intensity (Fig. 89c). In this case a deceptively simple 1 : 2 : 1 triplet results which appears to indicate that the protons are equally coupled to each phosphorus.

This behaviour is of use since it is found that in the majority of complexes the spin coupling between phosphorus atoms *trans* to one another is large while that between atoms which are *cis* is small. Thus *trans* pairs of methyl phosphines give rise to a methyl triplet whereas *cis* pairs of phosphines give

145

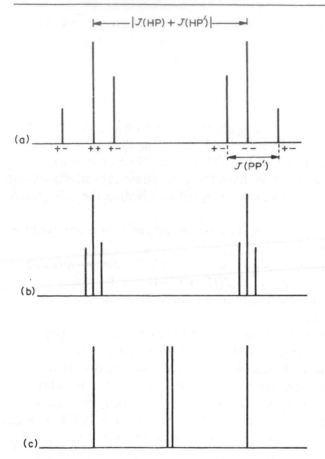

FIGURE 89

Proton spectra due to the two protons in the four spin system H . . . P . . . Metal . . . P′ . . . H′. In (b) $J(P\text{-}P') \ll | J(P\text{-}H) + J(P\text{-}H')|$ and the spectrum is essentially a doublet while in (c) $J(P\text{-}P') > | (J(P\text{-}H) + J(P\text{-}H')|$ and the spectrum approximates a 1 : 2 : 1 triplet. The small signs indicate the PP′ spin combinations giving rise to the different sub-spectra.

a methyl doublet. In complexes containing three phosphine ligands two may be *trans* but one must be *cis* and one should see both patterns. In Fig. 90 the proton spectrum is shown of the complex $[(C_6H_5)_2PMe]_4RuH_2$ which contains two *trans* phosphines and two *cis* phosphines. This is clearly indicated in the spectrum.

146

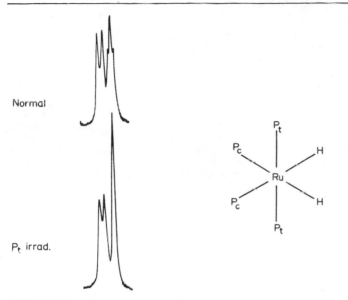

Normal

P$_t$ irrad.

FIGURE 90

Proton spectrum of the methyl groups and structure of the octahedral complex *cis* [(C$_6$H$_5$)$_2$ P CH$_3$]$_4$ RuH$_2$. The phosphine ligands are represented simply by the letter P. Two phosphines (P$_t$) are *trans* and give rise to the high field triplet in the upper trace while the remaining two (P$_c$) are *cis* and give rise to a doublet. The lower trace shows the effect of strongly double irradiating the *trans* phosphine phosphorus resonance (After Dewhurst, Keim, and Reilly).

In complexes which contain several different ligands, such as RhCl$_2$Et(CO) (PMe$_2$(C$_6$H$_5$))$_2$, there may also be effects due to the lack of symmetry. The complex structure is shown in Fig. 91. The two phosphines are *trans* and the phosphine methyl groups give triplets. However, because there is no plane of symmetry along the phosphine-rhodium bond, the methyl environments do not average to the same during a rotation of the phosphine and two triplets are observed. We have met this effect before with methylene protons attached to a carbon atom carrying three different substituents.

Metal complexes can also be formed with a variety of unsaturated compounds. The π-electrons of the unsaturated bonds form bonds to the metal and in so-doing their bond order is reduced. Olefinic protons for instance are normally descreened but when a π-complex is formed the bond anisotropy changes and we see a high field shift. Thus if 1, 4-dimethyl naphthaline is

147

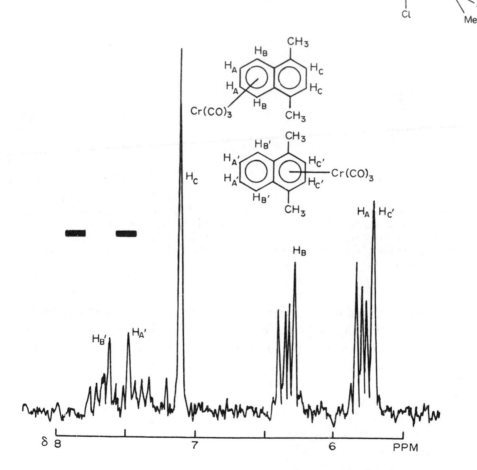

FIGURE 91
Structure of the complex $RhCl_2Et(CO)(PMe_2(C_6H_5))_2$. There is no plane of symmetry along the Rh-P bonds and so the P-methyl groups are never equivalent and give rise to two triplets instead of one. The symbol ϕ represents a phenyl group C_6H_5., the Me a methyl group CH_3 and the Et an ethyl group CH_3CH_2.

FIGURE 92
Proton spectrum of a mixture of the two complexes formed between 1, 4-dimethyl naphthalene and chromium tricarbonyl. The two sets of resonances are differentiated by primed letters. The two black bars indicate the region in which naphthalene resonates. (After Deubzer, Fritz, Kreiter, and Öfele.)

148

complexed with chromium tricarbonyl (Fig. 92) the chromium can bond into the π-system of either ring and two products are obtained. The spectrum of each consists of a methyl singlet (not shown) a singlet for the isolated, isochronous C protons at positions 2 and 3, and an $[AB]_2$ spectrum for the 5, 6, 7, and 8 protons. These latter two features can be clearly observed for both compounds in Fig. 92 where it will be seen that for the predominant complex the C singlet is low field of the $[AB]_2$ multiplet so that complexing has occurred *via* the unsubstituted ring, whereas in the minor component the C singlet is high field of the $[AB]_2$ multiplet, indicating complexing in the methyl substituted ring.

7.5   Variable temperature studies of some dynamic systems

Variable temperature nmr studies enable us to observe and measure the changes in rate of some dynamic process which affects the n.m.r. spectra. The first example here concerns the behaviour of aluminium trimethyl. This was believed to be dimeric Al_2Me_6 with two bridging and four terminal methyl groups but at room temperature its proton nmr spectrum is simply a sharp singlet suggesting that all six methyls are equivalent. This is contrary to what would be expected since the bridging methyls are in a different environment to the terminal methyl groups. In cyclohexane solution however it was found that as the temperature was reduced the methyl resonance broadened and separated into two singlets of intensity ratio 1 : 2, arising respectively from the bridging and terminal methyl groups. The two sorts of methyl group are thus interchanging roles rapidly on the n.m.r. time scale at room temperature. The activation energy for exchange was calculated from the exchange rates obtained from the line-shapes and was found to be 65 *KJ* mol^{-1}. This is probably similar to the energy of dissociation of the dimer in solution so that exchange may proceed via dissociation of the dimer. It was also possible to conclude from these spectra that the carbon atom of the bridging methyl groups probably forms the bridge link and that none of its hydrogens were directly involved, since these were all equivalent down to the lowest temperatures (Fig. 93).

A second field of study where nmr has been of assistance is that of σ-cyclopentadienyl metal complexes (Fig. 94a). The room temperature proton spectrum of these compounds is a singlet and so does not differentiate them from the π-cyclopentadienyl complexes where the metal is placed

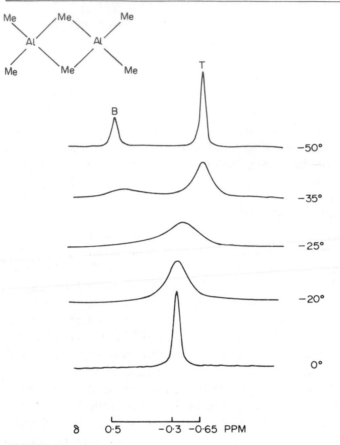

FIGURE 93

Proton spectra at several temperatures of aluminium trimethyl dimer dissolved in cychohexane. The terminal methyl groups (T) resonate high field of TMS but the involvement of the bridging methyls in further bonding results in their being descreened (b) (After Brownstein, Smith, Erlich, and Laubengayer.)

above a face of the ring and is equally bound to all five carbons, as in ferrocene. As the temperature is reduced however the spectrum of a σ bonded ring broadens and splits into three separate peaks corresponding to the three types of proton in the σ compound (Fig. 94b). This behaviour is of course indicative of rapid exchange at room temperature with the metal attached successively to each ring carbon atom thus averaging the proton environments to a single one. The exchange can take place in one of three ways: (i) by 1 : 2

150

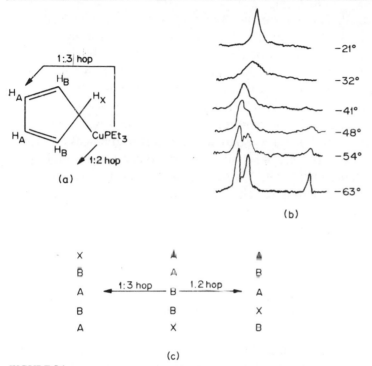

FIGURE 94

(a) A σ-cyclopentadienyl compound. The protons form an $[AB]_2 X$ system and should give rise to three resonances. (b) The proton spectrum at several temperatures of σ-cyclopentadienyl (triethyl phosphine) copper (I). The X proton is found to high field and the olefinic A and B protons to low field. Note the assymmetry in the low field region between −41° and −54°. (c) When a metal hop occurs the hydrogens interchange positions in different ways depending upon whether a 1 : 2 or a 1 : 3 hop is made.

hops, i.e. the metal moves always from one carbon atom to either of the adjacent two; (*ii*) by 1 : 3 hops where the metal always moves to either of the next but one carbons; or (*iii*) randomly with processes (*i*) and (*ii*) of equal probability. The spectra in Fig. 94b allow us to choose which process is occurring. Why this is so is best seen by reference to Fig. 94c. This indicates that when a 1 : 2 hop occurs an A spin becomes A, the other A spin becomes B, a B spin becomes A and so on. Now the A and B spins resonate close together so that when an A ↔ B interchange occurs there is only a small degree of frequency

151

uncertainty introduced. The X resonance however is well removed from both A and B so that when an A ↔ X or B ↔ X interchange occurs the frequency uncertainty, and consequent line broadening, is much greater. When an 1 : 2 hop occurs only B interchanges with X whereas when a 1 : 3 hop occurs only A interchanges with X. Thus in the case of 1 : 2 hops we expect the B resonance to be broader than the A resonance while if 1 : 3 hops occur then the A resonance will be the broader. If random hops occur the broadening will be equal. Reference to Fig. 94b shows that the A and B resonances are non-symmetrical, especially between $-41°$ and $-54°$ and that the random hop mechanism is immediately ruled out. Choice of the 1 : 2 or 1 : 3 mechanism depends upon correctly assigning which of the two low field resonances is A or B. Currently there is some controversy over this for the copper compound shown, but for the complex $(\sigma\text{-}C_5H_5)\,RuH(\pi\text{-}C_5H_5)$, which contains both a σ- and a π-bonded ring and whose σ-ring spectrum shows similar temperature dependence to the copper compound, it has been established from the structure observed in the resonances at the lowest temperature, which of the olefinic resonances couples most strongly to H_X and therefore which is H_B. The results indicate the 1 : 2 hops as the correct mechanism of exchange in this case.

The hydration complexes formed by ions undergo rapid exchange processes, in this case between molecules of the bulk solvent and of the hydration complex. The rates of exchange for the alkali metal salt solutions are too fast for nmr to be able to see separate resonances for bulk solvent and for solvation solvent, but for the smaller, more highly charged ions such as Be^{2+} and Al^{3+} the rates of exchange are much slower and the technique has made considerable contributions to our knowledge.

The greatest amount of data has been obtained for the aluminium-(III) ion. The hydration complex $Al(H_2O)_n^{3+}$ contains three magnetically active nuclei, ^1H, ^{17}O, and ^{27}Al, and all three have been used in a variety of studies concerned with determination of the value of n and of the rates of exchange of hydration water. Because of the hydrolysis reaction of the ion:

$$H_2O + Al(H_2O)_n^{3+} \rightleftharpoons Al(H_2O)_{n-1}(OH)^{2+} + H_3O^+$$

there was reason to believe that the water oxygen and hydrogens exchange at different rates so that ^1H and ^{17}O nmr might give different results.

When the ^{17}O resonance is observed (using ^{17}O enriched material) only a single line can be seen due to bulk and solvation water. If a paramagnetic salt is added with a cation having a short-lived hydration complex then the bulk water all rapidly comes under the influence of the ion and its unpaired electrons and suffers an average contact shift, whereas water bound to Al^{3+} does not. Separate bulk and hydration water resonances can then be observed. (Fig. 95).

$$\underset{^{17}\text{O normal chemical shift}}{H_2O^* \text{ (bulk)}} + \underset{^{17}\text{O contact shift}}{Co^{2+} (H_2O)_x} \rightleftharpoons H_2O + Co^{2+}(H_2O)_{x-1}\ (H_2O^*)$$

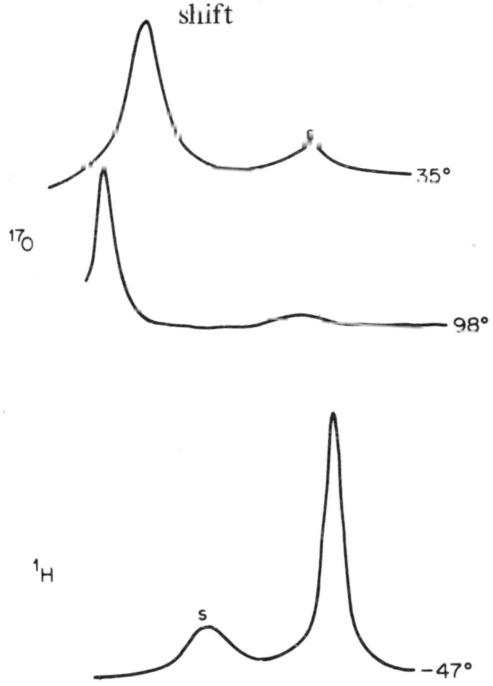

FIGURE 95
Upper: ^{17}O spectra of aqueous $AlCl_3$ with added Co^{2+}. The bulk water is shifted to low field. Raising the temperature increases the rate of oxygen exchange and broadens the resonance of the bound water. (After Fiat and Connick.) Lower: ^{1}H spectrum of concentrated (3M) aqueous $AlCl_3$ at −47° The water in the solvation complex is seen 4·2 ppm to low field of the bulk water. (After Schuster and Fratiello.) Comparison of the areas of the two peaks in each spectrum allows the hydration number to be calculated and is always near 6·0. S represents the resonance due to solvation water in each case.

Comparison of the areas under the two peaks, together with a knowledge of the water and aluminium content of the solution, allows the value of n to be calculated. This is invariably found to be near 6·0. Raising the temperature causes changes in line-shape which can be used to calculate rates of exchange. This gives the kinetic parameters for exchange:

rate constant k_O = $0·13 \, s^{-1}$ at 25°

ΔH^* = $113 \, kJ \, mol^{-1}$

ΔS^* = $+ 117 \, joule \, deg^{-1} \, mol^{-1}$

The positive sign of the entropy change indicates that the water exchanges by a dissociative mechanism.

The proton resonance of an aluminium chloride solution is also a singlet at room temperature. Fortunately the salt is very soluble so that its saturated solution has a considerably depressed freezing point. This allows the solution to be cooled to $-47°$ at which temperature separate resonances are seen for bulk and bound water protons. Variable temperature experiments enable the rate constant for proton exchange to be calculated and this is found to be, $k_H \approx 10^5 \, s^{-1}$ at 25°. Thus the exchange rates of oxygen and hydrogen differ by several orders of magnitude.

The ^{27}Al resonance of these solutions is a narrow singlet. This provides further proof that the ion is symmetrical around the aluminium, as would be the case for octahedral $Al(H_2O)_6^{3+}$. An unsymmetrical complex with n equal to 5 or 7 would have a line broadened by quadrupole relaxation.

7.6   Contact shifts and pseudo contact shifts

We have just seen how the contact shift phenomenon can be used to cause favourable chemical shifts in the study of aqueous salt solutions. Essentially we are using the large magnetic moment of the unpaired electron to produce the shift and are relying on rapid exchange to minimize the relaxation field of the electron at the water oxygens, which stay well removed from the ion for the major part of the time. Thus we avoid too much line broadening of the water resonance.

The contact shift phenomenon is also coming increasingly to be used to help simplify the spectra of certain organic compounds. A paramagnetic

154

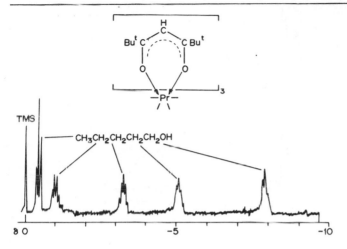

FIGURE 96
The paramagnetic complex tris (2,2,6,6-tetramethyl heptane-3, 5 dionato)-praseodymium has three
bulky tridentate ligands surrounding the metal in such a way that it is octahedrally coordinated by
oxygen. Only one of the ligands is shown above. Molecules with lone pairs of electrons also approach
the paramagnetic metal since the lanthanide can attain coordination numbers greater than six. Their
protons can thus interact with the unpaired electrons on the metal and undergo contact shifts. In the
case of n-pentanol a first-order spectrum is obtained (Bu^{t-} represents (CH$_3$)$_3$ C- i.e. the t-butyl group.
This type of ligand is used because it renders the ions soluble in organic solvents). (After Briggs, Frost,
Hart, Moss, and Staniforth.)

octahedral complex of a lanthanide element (Eu or Pr, Fig. 96) is dissolved
in a solution of the compound. The lanthanides are capable of assuming
higher coordination numbers than six so that if the organic molecule possesses
a suitable coordination site (e.g. O or N) it can interact with the complex. This
produces an average pseudo contact shift of the protons in the organic
molecule which can be very large. For instance the normal spectrum of
pentanol is found low field of TMS and consists of a triplet due to the
methylene protons adjacent to the alcohol group and two groups containing
a series of overlapping lines due to the remaining protons which all have
similar chemical shifts. If a praseodymium complex is added (Fig. 96) the
resonances all move upfield of TMS and each become separated sufficiently
to give first-order spectra. Alternatively if a europium complex is used the
contact shift is down field. These phenomena are by no means fully under-
stood but already form a useful tool to assist the organic chemist.

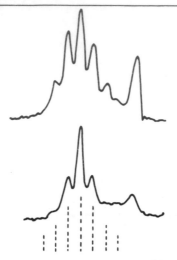

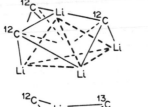

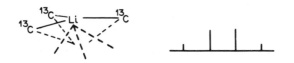

FIGURE 97

The ^{7}Li resonance of ^{13}C enriched methyl lithium at $-80°$. The high field singlet is a marker due to LiI. The upper trace is from a 50 per cent ^{13}C enriched sample and the lower is from a 25 per cent enriched sample. The latter contains more of the molecules with lithium adjacent to three ^{12}C atoms and so has the relatively more intense centre singlet. The stick diagram shows how the septet originates. (After McKeever, Waak, Doran, and Baker.)

156

7.7   The use of less common nuclear resonances

Because the proton is in many ways the most convenient nucleus to study it commonly features in nmr work. The preceding five sections of this chapter have been based almost entirely on the proton and the only other nuclei mentioned were ^{19}F, ^{17}O, and ^{27}Al. At the beginning of the book we saw that there are many more magnetically active nuclei than these and the purpose of this section is to survey a few of the results that have been obtained with other nuclei. Examples have been chosen from groups 1A, IIIB, VB, VIIB and a transition metal in group VIII only. We have already discussed ^{13}C (group IVB) and ^{17}O (group VIB) in previous pages. Group IIA has not been much studied.

7.7.1   The lithium-7 resonance

The lithium alkyls have been extensively studied using the ^7Li resonance ($I = \frac{3}{2}$). They undergo rapid exchange reactions at room temperature and resolved spectra are best obtained at reduced temperatures. A particularly striking example of this has been the use of ^7Li nmr to determine the structure of methyl lithium. This is a tetramer $(LiMe)_4$ but its structure was uncertain. The problem was solved by preparing samples containing 25 to 50 per cent ^{13}C enriched methyl groups. At $-80°$ ^7Li septets due to carbon-lithium spin coupling were resolved (Fig. 97). The structure of the tetramer is based on a tetrahedron of lithium atoms with the carbon atoms placed symmetrically above each face of the tetrahedron and bridging all three lithium atoms. Each lithium atom can spin couple equally to the three adjacent carbon atoms, $^1J\,(^7Li - {}^{13}C) = 15\,Hz$, and the ^7Li multiplicity depends upon the number of these which are the ^{13}C isotope. Thus three ^{12}C give a singlet, two ^{12}C and one ^{13}C a doublet, one ^{12}C and two ^{13}C a $1:2:1$ triplet and three ^{13}C a $1:4:4:1$ quartet. These multiplets interlace to give a septet whose line intensities depend upon the ^{13}C enrichment and with a separation equal to half the coupling constant. This is shown diagrammatically in Fig. 97 together with two ^7Li spectra of 25 and 50 per cent ^{13}C enriched materials.

7.7.2   The boron -11 and aluminium-27 resonances

The structure of diborane B_2H_6 was a problem which excited interest for many years and work with this compound led eventually to the concept of the multicentre bond with hydrogen acting as a bridging atom. The nmr spectra of diborane are entirely consistent with the double bridged structure. Fig. 98 shows spectra obtained recently using isotopically nearly pure $^{11}B_2H_6$ which gives a clearer pattern than diborane containing the normal 19·6 per cent of ^{10}B. The boron spectrum is basically a triplet of triplets arising from two isochronous boron atoms which are each coupled to two terminal protons and, by a smaller amount, to the two bridge protons. The proton spectrum consists basically of a 1 : 1 : 1 : 1 quartet to low field due to the terminal protons, which each have a major spin coupling to one boron atom, and a 1 : 2 : 3 : 4 : 3 : 2 : 1 septet of half the intensity to high field due to the bridge protons which are coupled equally to both boron atoms. The bridge protons also have a small spin coupling to all the terminal protons and each line of the septet is further split into a 1 : 4 : 6 : 4 : 1 quintet. Both the terminal proton quartet and the ^{11}B spectrum contain fine structure due to second-order spin coupling which arises because both the ^{11}B atoms and the terminal protons are not magnetically equivalent and the main peak separations are given by $^1J(B-H_t) + \,^3J(B-H_t)$. The spectrum can only be reconciled with the structure shown in the figure.

^{11}B nmr is also used extensively in the investigation of the higher polyhedral hydrides of boron and of the carboranes; compounds in which one or more of the boron atoms in a hydride have been replaced by carbon. Here we give one example of how the ^{11}B spectrum was used to determine the position of substitution of a dimethylphosphino group in pentaborane-9.

The parent hydride, B_5H_9 consists of a square-based pyramid of five boron atoms each carrying a hydrogen atom and with four more protons bridging the four basal borons. Its ^{11}B spectrum consists of two resonances in the intensity ratio 1 : 4 for the apical and basal borons respectively. The resonances are doublets due to spin coupling to the terminal hydrogen atoms. Coupling to the bridge hydrogens is not resolved.

FIGURE 98

The low temperature proton 100 MHz (a), and boron 19·25 MHz (b) spectra of diborane, B_2H_6, which was isotopically virtually pure in ^{11}B. The sharp vertical line in (a) is TMS. The terminal proton quartet is low field of this line and the bridge proton septet of quintets is to high field. The ^{11}B spectrum (b) is basically a triplet of triplets. Both this and (a) however contain second-order splittings. The figures around the formula give the various coupling constants. H_t are terminal and H_b are bridge hyrogen atoms respectively. (After Farrar. Johannensen, and Coyle.)

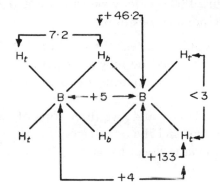

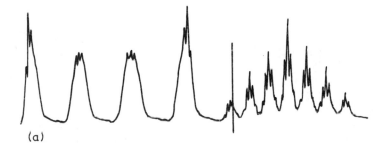

(a)

(b)

When a hydrogen is substituted by a PMe_2 group to give $B_5H_8PMe_2$ the substitution could occur in three possible positions:

(*i*)   On the apical boron. This would not affect the form of the ^{11}B spectrum greatly except that the boron-phosphorus coupling constant might differ from the boron-hydrogen value and the resonances might exibit chemical shifts.

(*ii*)   Terminally on a basal boron. This would give three sorts of basal boron, the substituted one, the two adjacent and the one diagonally opposite, in addition to the apical boron. We might expect to see four resonances in the

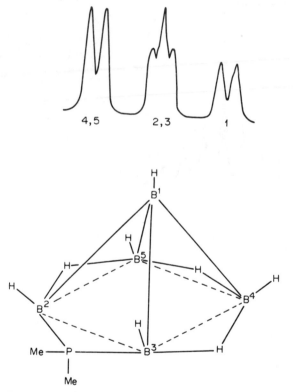

FIGURE 99

The ^{11}B spectrum of dimethylphosphinopentaborane. The numbers under the peaks refer to the numbered borons on the diagram below. The directly bonded coupling constant, ^1J(B-H), for the apical (B^1) and normal basal (B^{4,5}) borons is about 155 Hz. (After Burg and Heinen.)

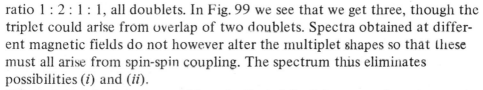

FIGURE 100
The ^{27}Al spectrum of a solution of anhydrous aluminium chloride in methyl cyanide. The low-field peak is due to $AlCl_4^-$ and the up-field peak to $Al(MeCN)_6^{3+}$. The chemical shift between the two resonances is 136 ppm. These spectra were obtained in the derivative mode. (After Hon.)

ratio $1 : 2 : 1 : 1$, all doublets. In Fig. 99 we see that we get three, though the triplet could arise from overlap of two doublets. Spectra obtained at different magnetic fields do not however alter the multiplet shapes so that these must all arise from spin-spin coupling. The spectrum thus eliminates possibilities (*i*) and (*ii*).

(*iii*) A bridge hydrogen could be substituted. In this case we have two sorts of basal boron, two bonded to both a bridge hydrogen and a bridge phosphorus and two bonded to two bridge hydrogens. We would thus expect three resonances in the ratio $2 : 2 : 1$, two due to basal and one due to apical boron. This is the spectrum observed. The presence of a $1 : 2 : 1$ triplet · indicates that two boron atoms are coupled to two spin $\frac{1}{2}$ nuclei roughly by an equal amount, i.e. to a terminal hydrogen and to a bridge phosphorus. This could not occur if substitution had been terminal.

The ^{27}Al resonance has been increasingly used in the last few years, especially in the study of solutions containing aluminium ions, Al^{3+}. Two examples are given. If anhydrous aluminium chloride is dissolved in methyl cyanide two ^{27}Al resonances are seen in the resulting solution (Fig. 100). The low-field resonance coincides with that known to arise from the anion $AlCl_4^-$ and it is therefore proposed that we get reaction with the solvent to

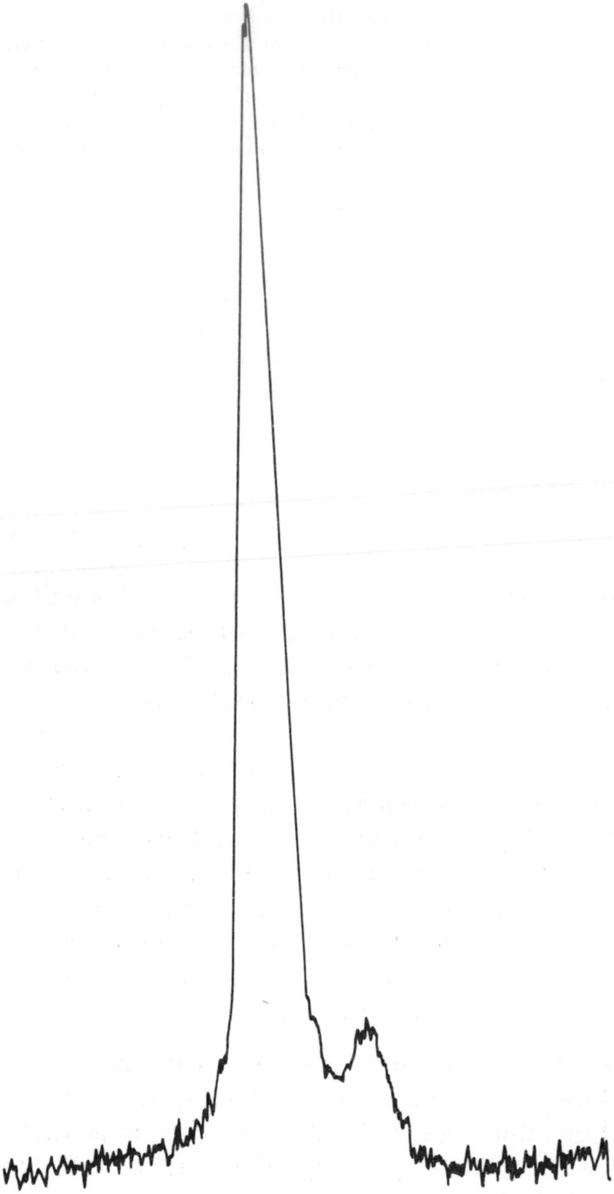

FIGURE 101
The 23·45 MHz ^{27}Al spectrum of aqueous aluminium sulphate solution. The large, low-field peak is due to $Al(H_2O)_6^{3+}$ and the small high-field peak to a sulphate complex, probably $Al(H_2O)_5 (SO_4)^+$. The peaks are separated by 77 Hz.

form this anion and a solvated cation.

$$4 \, AlCl_3 + 6 \, MeCN \rightleftharpoons 3 \, AlCl_4^- + Al(MeCN)_6^{3+}$$

This picture is confirmed by proton spectroscopy which detects separate signals for bulk and bound solvent. Comparison of the peak areas indicates the presence of an average of 1·5 bound MeCN molecules to every aluminium atom.

An aqueous aluminium sulphate solution also gives two ^{27}Al resonances, a major one due to $Al(H_2O)_6^{3+}$ and a minor one due to a complex in which one water molecule has been replaced by a sulphate ion. Certainly the complex ion must remain six coordinate since its resonance is hardly any broader than that of the hexaaquo ion (Fig. 101).

7,7,3.   A transition metal, the cobalt-59 resonance

^{59}Co nmr has had a limited application in the study of cobalt complexes The nucleus has a large quadrupole moment and so its lines are in general broad. Fortunately its chemical shifts are large so that the broad lines do not obscure changes in environment. The shifts extend over at least 13 000 ppm ($K_3Co(CN)_6$ = 0, $K_3Co(C_2O_4)_3$ = 13 000) and are thought to arise from changes in the paramagnetic screening mechanism. The lines due to symmetrical complexes are of course the least broad and in one case fine structure has been observed (Fig. 102). A spectrum is shown due to $Co(PF_3)_4^-$ which exhibits quintet splitting due to coupling to four phosphorus atoms $^1J(^{59}Co\text{-}^{31}P)$ = 1222 Hz and to twelve fluorine atoms $^2J(^{59}Co\text{-}^{19}F)$ = 57 Hz. The resonance occurs 4220 ppm from $Co(CN)_6^{3-}$.

7.7.4   Phosphorus-31 n.m.r.

There is an extenisve literature of phosphorus nmr. The ^{31}P isotope has an abundance of 100 per cent and a spin of $\frac{1}{2}$ so that it gives narrow lines. It

FIGURE 102

The ^{59}Co resonance of the $Co(PF_3)_4^-$ ion in aqueous solution. The ion is tetrahedral with low electric field gradient at the cobalt nucleus so that the lines are correspondingly narrow and spin coupling can be observed to phosphorus and to fluorine. Spectra were run in the derivative mode. (After Lucken, Noack, and Williams.)

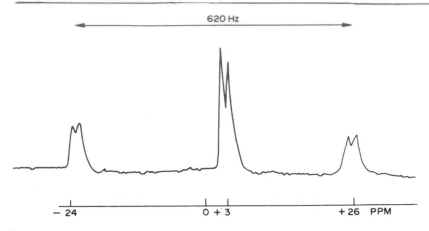

FIGURE 103

Phosphorus-31 spectrum of the isohypophosphate anion at 12·3 MHz. The external standard was 85 per cent H_3PO_4. The small splitting on each of the three peaks is 17 Hz. (After Van Wazer, Callis Shoolery, and Anderson.)

behaves rather like the proton except that its signal is appreciably weaker and its chemical shifts and spin coupling constants tend to be larger. Chemical shifts are to some extent diagnostic of the coordination number (3, 4, or 5) of the phosphorus though overlap of each region occurs. An example of a ^{31}P resonance has already been given (Fig. 54) and many studies have been made of other organo-phosphorus compounds. The nucleus has also been used to study the numerous phosphorus oxy-acids and has helped considerably to elucidate their structure. Thus the spectrum of the isohypo-phosphate anion, $HP_2O_6^-$, is shown in Fig. 103. The spectrum consists of a doublet of intensity 1 and two smaller doublets of intensity $\frac{1}{2}$. It is not a 1 : 2 : 1 triplet since the centre doublet is displaced from the exact centre of the small doublets. There are thus two non-equivalent phosphorus atoms in the molecule. They can be expected to spin couple and this explains the mutual small doublet splitting of 17 Hz. One phosphorus atom resonance is further split by the very large coupling constant of 620 Hz. This splitting has no counterpart in the phosphorus spectrum and so must arise from coupling to another type of nucleus, i.e. to the proton in the molecule. The large value of the coupling constant is diagnostic of a directly bonded interaction and so we can write the formula as:

$$\left[\begin{array}{cc} O & O \\ | & | \\ O-P-O-P-O \\ | & | \\ O & H \end{array}\right]$$

The fact that coupling to the proton is observed at all indicates that it is a non-exchanging proton and not an acidic OH proton.

The neutralization of phosphoric acids can also be followed by nmr since both the chemical shift of the phosphorus and its coupling constants in the polyphosphoric acids are sensitive to the state of ionization of the acid. If we add alkali to an orthophosphoric acid solution a single resonance is observed at all pH values because of rapid exchange between the various ions present.

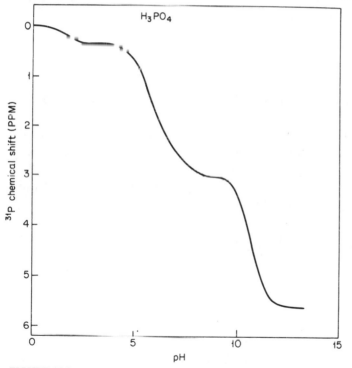

FIGURE 104

Phosphorus-31 chemical shifts of orthophosphoric acid, H_3PO_4, at different degrees of neutralization. At pH 0 we have almost entirely unionised H_3PO_4, at pH 3 we have $H_2PO_4^-$, at pH 9 we have HPO_4^{2-}, and above pH 12 we have the fully ionised PO_4^{3-}. (After Crutchfield, Callis, Irani, and Roth.)

165

The chemical shift however changes with pH as the proportion of each ionic species alters. Three plateaux are observed in the titration curve of Fig. 104 corresponding to regions where we have predominantly $H_2PO_4^-$, HPO_4^{2-} and PO_4^{3-}.

7.7.5 Fluorine-19 n.m.r.

The fluorine-19 nucleus is also an important one since it is 100 per cent abundant and gives a signal intensity second only to that of the proton. It is a spin $\frac{1}{2}$ nucleus and its spectroscopy is very similar to that of the proton except that the chemical shifts between fluorine atoms in different environments in a molecule are often very large. If spin coupling exists then first-order patterns are obtained. We have already seen an example of this in Section 7.4 on the fluorotitanate complexes. Two further examples of fluorine n.m.r. spectroscopy are now given.

The first concerns the tetrafluoroborate anion BF_4^-. The fluorine atoms are arranged tetrahedrally around the boron so that the electric field gradient at the boron nucleus is small and electric quadrupole relaxation is slight. Thus the boron spin coupling to the fluorine atoms can be observed in the fluorine spectrum. This consists of a well-resolved quartet due to the 80·6 per cent of $^{11}BF_4^-$ ions (boron-11, $I = \frac{3}{2}$) and a septet due to the 19·4 per cent of $^{10}BF_4^-$ (boron-10, $I = 3$) (Fig. 105). The spectrum is particularly interesting in that it clearly demonstrates that the fluorine atoms attached to the boron isotope ^{11}B are chemically shifted from those attached to ^{10}B. This is called an isotope shift and arises because the vibrational states of the two ions differ slightly so that the electron distributions within the bonds are also different.

The coupling constants to the two isotopes should be in the ratio of the magnetogyric ratios which is $\gamma(^{10}B)/\gamma(^{11}B) = 0·335$. The corresponding coupling constants are found to be in the ratio 0·33.

The ion is unusual in that the coupling constant varies over a considerable range depending on its environment. In aqueous NH_4BF_4, $^1J(^{11}B-^{19}F)$ is 1·2 Hz whereas in $NaBF_4$ solution it may attain values of up to 5·0 Hz, depending upon the concentration. In non-aqueous solvents the coupling constant may change sign. This behaviour is thought to arise because of the presence of two large electronic contributions to the coupling constant which are opposite in sign and therefore nearly cancel. Changes in environment produce small but different changes in each contribution which are large in relation to the

166

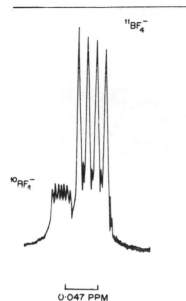

$^{11}BF_4^-$

$^{10}BF_4^-$

0·047 PPM

FIGURE 105
Fluorine-19 spectrum of the BF_4^- ion in an aqueous solution of ammonium tetrafluoroborate. The septet at low field arises from fluorine bonded to the isotope ^{10}B ($I = 3$) and the quartet at high field from fluorine bonded to ^{11}B ($I = \frac{3}{2}$). The different isotopic masses cause a fluorine chemical shift of 0·047 ppm. The coupling constants are $^1J(^{11}B{-}^{19}F) = 1·2$ Hz; $^1J(^{10}B{-}^{19}F) = 0·4$ Hz.

net coupling constant and so become particularly obvious.

As a second example of fluorine spectroscopy we consider the sterochemistry of the substituted six-membered hetero cyclic phosphonitrilic trimers (Fig. 106). This is capable of considerable elaboration if several different substituents are introduced into the molecule. The two molecules shown are isomers and differ only in that in one the dimethylamino groups are placed on the same side of the P_3N_3 ring (the *cis* isomer) while in the other they are on opposite sides (the *trans* isomer). The fluorine spectra differentiate the two molecules. In the *cis* isomer one of the geminal fluorines, F_A, is on the same side of the ring as the two dimethylamino groups and F_B is on the same side as the two chloro substituents. The fluorine atoms have different chemical environments and so have different chemical shifts. They can couple with each other and with the directly bonded phosphorus atom. As a result we see two large doublets $^1J(P{-}F) = 905$ Hz with each line split into a small doublet $^2J(F_A{-}F_B) = 75$ Hz. The line intensities are perturbed and the spectrum can also be considered as a pair of AB sub-spectra, i.e. as part of an ABX system.

In the *trans* isomer both sides of the ring are identical as far as the fluorine atoms are concerned and a single doublet is obtained due to coupling to the

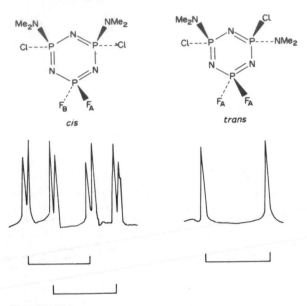

FIGURE 106

Fluorine-19 spectra of *cis* and *trans* 1, 3-bis dimethyl-1, 3-dichloro-5, 5-difluoro triphosphorinitrile. The dashed lines on the formulae indicate groups behind the plane of the paper and the solid lines those in fromt. The markers under the spectra refer to the directly bonded ^1J(P-F) coupling. (After Green and Sowerby.)

phosphorus atom ^1J(P-F) $=$ 900 Hz

Further examples of fluorine spectroscopy will be found in the exercises at the end of the chapter.

7.8   N.M.R. of solids

The study of solid systems is usually omitted from most text books since the interest of chemists lies overwhelmingly with the high resolution study of liquid samples. Sufficient work of chemical interest has been carried out however to make it worthwhile to mention some of the possibilities.

We have already seen that the nmr resonances of solids are broad due to the direct through-space spin-spin interaction. The magnitude of the interaction, and therefore the line-width and line-shape, depends upon the distance between interacting spins and the number in an interacting cluster. Thus in many crystalline hydrates it is possible to distinguish whether the water of crystallization is present as H_2O, H_3O^+, or OH^-. For instance borax,

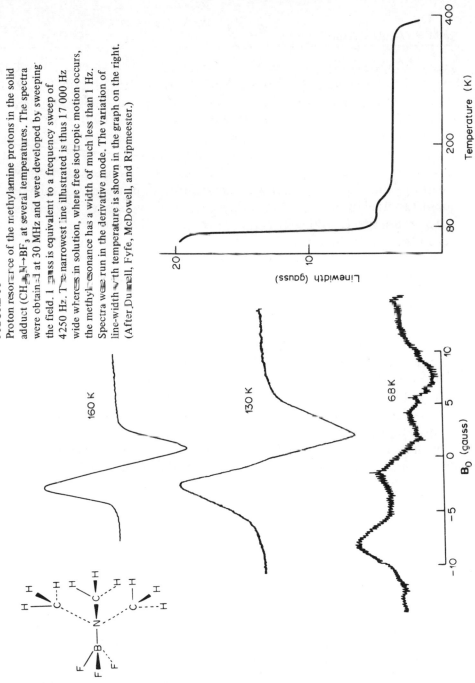

FIGURE 10⁻

Proton resonance of the methylamine protons in the solid adduct (CH₃N→BF₃ at several temperatures. The spectra were obtained at 30 MHz and were developed by sweeping the field. 1 gauss is equivalent to a frequency sweep of 4250 Hz. The narrowest line illustrated is thus 17 000 Hz wide whereas in solution, where free isotropic motion occurs, the methyl resonance has a width of much less than 1 Hz. Spectra were run in the derivative mode. The variation of line-width with temperature is shown in the graph on the right. (After Dunnell, Fyfe, McDowell, and Ripmeester.)

160 K

130 K

68 K

B₀ (gauss)

Linewidth (gauss)

Temperature (K)

169

$Na_2O \cdot 2B_2O_3 \cdot 10H_2O$ has been shown to have the structure $Na_2[B_4O_5(OH)_4]$ $8H_2O$, a hydrated scandium nitrate, $Sc_4O_3(NO_3)_6 \cdot 7H_2O$ should be formulated $[Sc_4O(OH)_4]$ $(NO_3)_6 \cdot 5H_2O$, while a hydrated form of gallium sulphate contains both hydronium ion and hydroxide ions and is $(H_3O)Ga_3(OH)_6(SO_4)_2$.

If the interacting spins are moving so that their relative orientations change, e.g. by rotation of the group of atoms around some axis, then their through space coupling and the line-width of their resonance is reduced. Thus by measuring line-widths at a series of temperatures, information can be obtained about the state of motion of various groups in a crystal. Fig. 107 shows the solid state proton spectra obtained from the complex adduct of trimethylamine and boron trifluoride, $(CH_3)_3N{\rightarrow}BF_3$. At $68°$ K the triplet shape of the spectrum is that expected for a triangular group of three neighbouring, stationary protons. As the temperature is increased to $103°$ K a marked narrowing occurs and the structure disappears from the resonance. This transition marks the onset of rotation of the methyl groups around the C-N bond. Further narrowing occurs between $100°$ K and $150°$ K and this is due to the onset of rotation of the whole $N(CH_3)_3$ molecule around the B-N bond. Finally just below $400°$ K the line narrows to a fraction of a gauss as the whole molecule starts to·rotate isotropically and to diffuse within the still *solid* crystal.

The fluorine resonance of the BF_3 part of the molecule is broad only below $77°$ K and the BF_3 group rotates around the B-N axis at all higher temperatures.

A similar study using the ^7Li resonance has detected the onset of diffusion of lithium in a lithium borate glass at $293°$ K (Fig. 108).

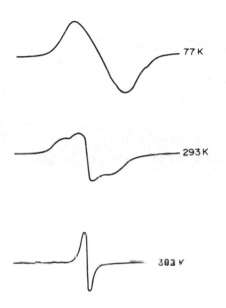

FIGURE 108
^{7}Li resonance in the lithium borate glass of composition 60/40 mole ratio Li_2O/B_2O_3. As the temperature is increased a proportion of the lithium ions start to diffuse and the resonance has a broad and a narrow component. At 382°K only the narrow component remains and its width suggests that the lithium is diffusing over only a limited volume. (After Bishop and Bray.)

7.9   Further exercises

1   The proton resonance of chloroform, $CHCl_3$, was measured as 730 Hz downfield of TMS. The spectrometer operating frequency was 100 MHz. What is the chemical shift of chloroform on the δ and τ scales? What would be the separation of the two resonances in Hz for a spectrometer operating at 60 MHz?

2   Fig. 18 illustrates an ethyl group spectrum. Measure the chemical shifts (δ) of the methyl and methylene protons and the coupling constant between them.

3   A sketch of the 40 MHz proton resonance of tetraethyl lead is shown in Fig. 109, with the relative positions of each line marked in Hz.(assuming first-order line positions). Lead contains the following isotopes:

^{206}Pb natural abundance 24% $I = 0$
^{207}Pb natural abundance 22% $I = \frac{1}{2}$
^{208}Pb natural abundance 54% $I = 0$

Explain the variuos features of the spectrum and calculate the coupling constants between lead and the methyl and methylene protons and also the interproton coupling constant. What is the chemical shift between the methyl and methylene protons?

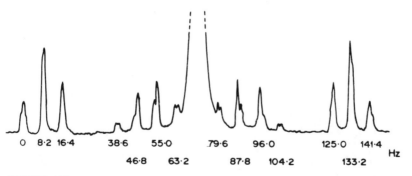

FIGURE 109
40 MHz proton spectrum of tetraethyl lead. (After Narasimhan and Rogers.)

The satellite line intensities are perturbed from the theoretical so that the spectrum is slightly second order. This is also evident in the tendency for extra splittings to appear in the quartet-triplet patterns. The effect is nevertheless only slight and can be shown to conform to the expected pattern for a separation equal to that between a methyl triplet one one side of the central resonance and a methylene quartet on the opposite side. What does this tell us about the relative signs of $^2J(^{207}Pb-^1H)$ and $^3J(^{207}Pb-^1H)$? Suggest why $^3J > {}^2J$.

4 Fig. 27c shows the spectrum of ascaridole. The spectrum features a closely coupled AB quartet at about $\delta = 6\cdot45$ ppm. The line positions, starting at the lowest field, are $395\cdot5$, $386\cdot9$, $385\cdot5$, and $376\cdot9$ Hz from TMS. Calculate $^3J(H_c\text{-}H_d)$ and the chemical shift between H_c and H_d in Hz.

5 Fig. 5J shows the proton spectrum of pentadeutero dimethylsulphoxide, $CD_3SO\,CD_2H$. The two deuterons split the proton resonance into a quintet. What should the relative intensities of the five lines be? What is $^2J(H\text{-}H)$ in the methyl group, given $^2J(H\text{-}D) = 1\cdot85$ Hz.

6 Construct a stick diagram for the fluorine and borane proton resonances of the complex adduct $HF_2P\cdot BH_3$ given that $^1J(^{11}B\text{-}H) - 103$ Hz, $^3J(F\text{-}H) = 26$ Hz, $^2J(P\text{-}H) = 17\cdot5$ Hz, $^3J(H\text{-}H) = 4\cdot0$ Hz, $^1J(P\text{-}F) = 1151$ Hz, and $^2J(H\text{-}F) = 54\cdot5$ Hz.

7 The proton resonances of TMS and cyclohexane are separated by $1\cdot43$ ppm the cyclohexane being to low field. If the TMS resonates at 60 000 000 Hz in a particular field locked spectrometer, at what frequency will the cyclohexane resonate?

8 The fluorine resonance of the hexafluoroniobate ion $^{93}NbF_6^-$ observed as a viscous solution in dimethylformamide is a well-resolved multiplet at $60°C$. The spin of ^{93}Nb is 9/2. How many lines should be observed in the multiplet? If the sample is cooled to $0°C$ or heated to $125°C$ the multiplicity is lost and a broad unresolved resonance is obtained. Explain these observations.

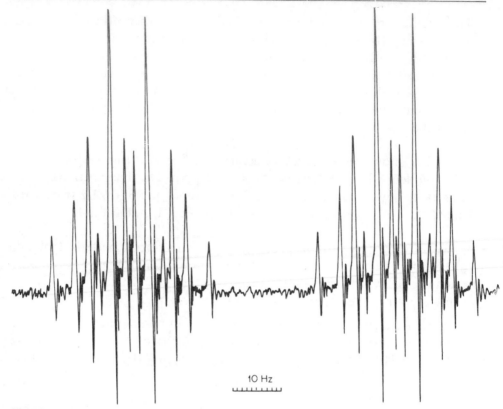

FIGURE 110
^{19}F resonance of the CF_2H group of $CF_3CF_2CF_2H$.

9 Fig. 110 shows the fluorine-19 resonance of the CF_2H group of 1, 1, 1, 2, 2, 3, 3-heptafluoropropane, $CF_3CF_2CF_2H$. The two fluorines are equivalent and are coupled to all the other magnetically active nuclei in the sample. Pick out the various multiplet patterns and measure the coupling constants $^2J(F\text{-}H)$, $^3J(F\text{-}F)$, and $^4J(F\text{-}F)$.

10 The proton spectrum of a sample of neat methanol, CH_3OH consists at 20° C of two singlets with the hydroxyl resonance 1·6 ppm low field of the methyl resonance. If the temperature is progressively reduced the singlets broaden; structure appears and at $-60°$ C the hydroxyl resonance is a 1 : 3 :

174

3 : 1 quartet and the methyl resonance is a doublet. In addition the hydroxyl resonance moves downfield and is now 2·3 ppm from the methyl doublet. Explain these observations.

11   Fig. 20e shows the methyl proton spectrum of $[Me_2NBCl_2]_2$ and ascribes it to spin spin coupling to two ^{11}B nuclei. The presence of 19·6 per cent of ^{10}B in the molecules is said to lead to line broadening. About 39 per cent of the molecules will contain one ^{10}B atom and one ^{11}B atom. Construct a stick diagram for the methyl resonances of these molecules and show how this reduces the resolution of the all ^{11}B septet. The spin of ^{11}B is $\frac{3}{2}$ and of ^{10}B is 3.

Answers

1    $\delta = 7\cdot3$ ppm $\tau = 2\cdot7$ 438 Hz

2   Chemical shifts are 1·48 and 3·56 ppm
       ^3J(H-H) $= 7$ Hz

3   ^3J(iI-H) $= 8\cdot2$ Hz ^3J(Pb-H) $= 125\cdot0$ Hz ^2J(Pb-H) $= 41\cdot0$ Hz
       methyl-methylene shift $= 0\cdot0175$ ppm
       methylene protons to high field

4    $J = 8\cdot6$ Hz, shift $= 5\cdot1$ Hz

5    $1:2:3:2:1$.

6   The proton resonance contains 48 lines and the fluorine resonance,
     where some lines overlap, contains only 12 lines. The spectra are
     illustrated in the *Journal of the Americal Chemical Society,* Volume 89,
     pages 1622–1623.

7   The resonances are separated by 85·8 Hz at 60 MHz. The cyclohexane
     protons are less screened, experience the larger applied magnetic field
     and so always precess at a higher frequency than do the protons in TMS.

8   The multiplicity is ten. The loss of resolution at one extreme is due to
     quadrupole relaxation and at the other extreme to exchange of fluoride
     ion, involving Nb-F bond breaking.

9   The pattern is a doublet of quartets of triplets with coupling constants
     of 53 Hz, 7·3 Hz, and 4·7 Hz. No lines overlap.

10   The hydroxyl proton is exchanging. Its chemical shift is determined by
     the degree of hydrogen bonding it experiences.

11   A $1:1:1:2:2:2:3:2:2:3:2:2:2:1:1:1$ multiplet is obtained.
     None of the lines coincide with those of the all ^{11}B multiplet.

Bibliography

Because of the importance of n.m.r. spectroscopy a considerable number of books, articles, and reviews on the subject have been published within the last twenty years. The following list represents only a small proportion of the total but is sufficient to give an entry into the n.m.r. literature.

Four comprehensive works are available for general reference.

Pople, J. A., Schneider, W. G., and Bernstein, H. J. (1959), *High resolution Nuclear Magnetic Resonance*, McGraw-Hill, New York.

Emsley, J. W., Feeney, J., and Sutcliffe, L. H. (1965), *High Resolution Nuclear Magnetic Resonance Spectroscopy* in two volumes; Pergamon Press, Oxford.

Jackman, C. M., and Sternhell, S. (1969), *Applications of Nuclear Magnetic Resonance Spectroscopy in Organic Chemistry* 2nd Edn. In the International Series of Monographs in Organic Chemistry Volume 5, Pergamon Press, Oxford.

Abragam, A., (1961), *The Principles of Nuclear Magnetism*, Oxford.

Several books have been published which contain a simple introduction to n.m.r. together with a variety of problems involving interpretation of spectral traces. Two such books are listed and the student may find these or other similar in the library.

Scheinmann, F., Ed. (1971), *An Introduction to Spectroscopic Methods for the Identification of Organic Compounds*, Vol. 1, *NMR and IR*, Pergamon Press, Oxford.

Dyer, J. R., (1965), *Applications of Absorption Spectroscopy of Organic Compounds*, Prentice Hall, London.

Two works dealing in more detail with theoretical considerations, particularly analysis of second-order spectra and which can be read as an introduction to some of the more comprehensive works are:

Roberts, J. D., (1969), *An Introduction to the Analysis of Spin-Spin Splitting in NMR*, Benjamin.

Lynded-Bell, R. M., and Harris, R. K., (1969), *Nuclear Magnetic Resonance Spectroscopy*, Nelson, London.

A short article in *Chemistry and Engineering News* on nuclear magnetic resonance should prove useful to the undergraduate and contains many examples of interest.

Bovey, F. A., (August 30, 1965), *Chemistry and Engineering News*, pp. 98–120.

The student may also care to read the following few original short papers which summarise the early and unexpected results which heralded the development of nmr as a subject useful to chemists.

Dickinson, W. C., (1950), *Phys. Revs.*, **77**, 736, *Observed chemical shifts in fluorine compounds and noted the effect of exchange.*

Proctor, W. G., and Yu, F. C., (1950), *Phys. Revs.*, **77**, 717, *^{14}N chemical shift between NH_4^+ and NO_3^-.*

Arnold, J. T., Dharmatti, S. S., and Packard, M. E., (1951), *J. Chem. Phys.*, **19**, 507, *First observation of chemical shifts in a single chemical compound.*

Gutowsky, H. S., and McCall, D. W., (1951), *Phys. Revs.*, **82**, 748, *An early observation of spin-spin coupling.*

Gutowsky, H. S., McCall, D. W., and Slichter, C. P., (1951), *Phys. Revs.*, **84**, 589, *A theory of spin-spin coupling.*

It should be remembered in reading the last two papers that the Hz separation of the ^{31}P doublet and ^{19}F doublet are the same. The gauss separation can be calculated from $\Delta B_0 = (J/\nu_0)B_0$ and is greater for ^{31}P since ν_0 is smaller for the fixed field used.

Index